The Emerging Role of Geomedia in the Environmental Humanities

ENVIRONMENT AND SOCIETY

Series Editor

Douglas Vakoch

As scholars examine the environmental challenges facing humanity, they increasingly recognize that solutions require a focus on the human causes and consequences of these threats, and not merely a focus on the scientific and technical issues. To meet this need, the Environment and Society series explores a broad range of topics in environmental studies from the perspectives of the social sciences and humanities. Books in this series help the reader understand contemporary environmental concerns, while offering concrete steps to address these problems.

Books in this series include both monographs and edited volumes that are grounded in the realities of ecological issues identified by the natural sciences. Our authors and contributors come from disciplines including but not limited to anthropology, architecture, area studies, communication studies, economics, ethics, gender studies, geography, history, law, pedagogy, philosophy, political science, psychology, religious studies, sociology, and theology. To foster a constructive dialogue between these researchers and environmental scientists, the Environment and Society series publishes work that is relevant to those engaged in environmental studies, while also being of interest to scholars from the author's primary discipline.

Recent Titles in the Series

The Emerging Role of Geomedia in the Environmental Humanities, edited by Mark Terry and Michael G. Hewson
The Bangladesh Environmental Humanities Reader: Environmental Justice, Developmental Victimhood, and Resistance, by Samina Luthfa, Mohammad Tanzimuddin Khan, and Munasir Kamal
Loren Eiseley's Writing across the Nature and Culture Divide, by Qianqian Cheng
The Saving Grace of America's Green Jeremiad, by John Gatta
Art and Nuclear Power: The Role of Culture in the Environmental Debate, by Anna Volkmar

Contesting Extinctions: Decolonial and Regenerative Futures, edited by Luis I. Prádanos, Ilaria Tabusso Marcyan, Suzanne McCullagh, and Catherine Wagner

Embodied Memories, Embedded Healing: New Ecological Perspectives from East Asia, edited by Xinmin Liu and Peter I-min Huang

Ecomobilities: Driving the Anthropocene in Popular Cinema, by Michael W. Pesses

Global Capitalism and Climate Change: The Need for an Alternative World System, Second Edition, by Hans A. Baer

Ecological Solidarity and the Kurdish Freedom Movement: Thought, Practice, Challenges, and Opportunities, edited by Stephen E. Hunt

Wetlands and Western Cultures: Denigration to Conservation, by Rod Giblett

Sustainable Engineering for Life Tomorrow, edited by Jacqueline A. Stagner and David S. K. Ting

Nuclear Weapons and the Environment: An Ecological Case for Nonproliferation, by John Perry

The Emerging Role of Geomedia in the Environmental Humanities

Edited by

Mark Terry and Michael Hewson

LEXINGTON BOOKS
Lanham • Boulder • New York • London

Published by Lexington Books
An imprint of The Rowman & Littlefield Publishing Group, Inc.
4501 Forbes Boulevard, Suite 200, Lanham, Maryland 20706
www.rowman.com

86-90 Paul Street, London, EC2A 4NE

Copyright © 2022 The Rowman & Littlefield Publishing Group, Inc.

All rights reserved. No part of this book may be reproduced in any form or by any electronic or mechanical means, including information storage and retrieval systems, without written permission from the publisher, except by a reviewer who may quote passages in a review.

British Library Cataloguing in Publication Information Available

Library of Congress Cataloging-in-Publication Data

Names: Terry, Mark (Film producer), editor. | Hewson, Michael (Lecturer in geography), editor.
Title: The emerging role of geomedia in the environmental humanities / edited by Mark Terry and Michael Hewson.
Description: Lanham : Lexington Books, [2022] | Series: Environment and society | Includes bibliographical references. | Summary: "This book provides the latest scholarship on the various methods and approaches being used by environmental humanists to incorporate geomedia into their research and analyses, examining how these new methodologies impact the production of knowledge in this field of study and promoting the impact of First Nation people perspectives"— Provided by publisher.
Identifiers: LCCN 2022031881 (print) | LCCN 2022031882 (ebook) | ISBN 9781666913422 (cloth) | ISBN 9781666913439 (ebook)
Subjects: LCSH: Geographic information systems—Social aspects—Case studies. | Environmental management—Data processing—Citizen participation—Case studies. | Indigenous data sovereignty—Case studies. | Human ecology and the humanities—Case studies. | Ecocriticism. | Digital divide.
Classification: LCC G70.212 .E436 2022 (print) | LCC G70.212 (ebook) | DDC 304.20285—dc23/eng20220919
LC record available at https://lccn.loc.gov/2022031881
LC ebook record available at https://lccn.loc.gov/2022031882

Contents

Acknowledgments ix

Introduction 1

Chapter One: GIS and the Environmental Humanities: How Citizen Scientists, Civil Servants, and Researchers Are Teaming Up to Study and Solve Environmental Issues 9
Mark Terry

Chapter Two: Wadawurrung Dja: The Ethnography and Biogeography of Pre-Colonization Wadawurrung Country in a Digital Realm 27
Susan Ryan, David S. Jones, Murray Herron, and Phillip Roös

Chapter Three: Tech for TEK: The Value of GIS Systems in Sustainable Community Planning and Indigenous Land Protection Initiatives 49
Shahreen Shehwar

Chapter Four: The Use of GIS by Indigenous Peoples in Charting Culture, Claims, and Country 63
Jigme Lhamo Tsering

Chapter Five: Ecofeminist Visualization: Reading GIS as a Bridge to Gendered Water Management in India: Reading GIS as a Bridge to Gendered Water Management in India 75
Pamela Carralero

Chapter Six: Ecologies of the Digital Map: GIS and the Geography of Autopoietic Worlding 93
Erik Tate

Chapter Seven: In the Retelling: Exploring Spatial Data as
    Narratives of Place   115
    Michael Hewson

Chapter Eight: Geomedia as a Pedagogical Tool: Toward
    Sustainability Competence: Toward Sustainability Competence   135
    Michael John Long

Chapter Nine: When Place Is Elsewhere: Pedagogy of Place for
    Planetary Health Education in a Digital Space   151
    Netta Kornberg

Chapter Ten: Geomedia in the Classroom: A Pedagogical Approach
    to GIS-Enhanced Ecocriticism   171
    Mark Terry, Erik Tate, and Shahreen Shehwar

Conclusion   185

Index   187

About the Contributors   193

Acknowledgments

Any anthology is a collaborative effort, but aside from the obvious chapter contributors and editors, there are also many behind the scenes who are instrumental in advancing this work to its final and published state. We, the editors, would like to acknowledge those supportive and influential individuals in this section.

OUR FAMILY, FRIENDS, AND COLLEAGUES

For Mark Terry (Family)

My patient, understanding, supportive, encouraging, and loving children: Herb and Mary Anne and their respective partners, Melissa and Ricardo; my wise and compassionate brothers, Hebé (Herb) and Chip (Bob), and their respective wives, Miriam and Sharon, as well as all their children and grandchildren, too numerous to name here.

For Mark Terry (Friends)

Among my friends acknowledged here are those who provided moral support, guidance, encouragement, and assurances, and who were always there for me when I needed a break. Your patience and understanding through this process is greatly appreciated and indirectly contributed to the completion of this book: Laura Bannon, Melanie Martyn, Michael Khashmanian, Janice Mennie, Stavros Stavrides, Carolyn Kelly, Blair Minnes, Michael Beighton, Gigi Stafne, Britta Schewe, Jackie Pearce, John McNamara, Timothy Hudson, Shirley Gossack, Sarwar Abdullah, Maeva Gauthier, Philip Jackson, Jon McSpadden, Tony Morrone, and Conor Fitzgerald.

For Mark Terry (Academic Colleagues)

Markus Reisenleitner; Susan Ingram, Martin Bunch, Sarah Flicker, Charles Hopkins, Katrin Kohl, Alice Hovorka, Jeff Grishow, Russell Kilbourn, James Orbinski, Theresa Dinh, Shana Yael Shubs, the Dahdaleh Institute for Global Health Research, the Royal Society of Canada, York University, and Wilfrid Laurier University. Special thanks to Courtney Morales, Kasey Beduhn, and Emma Ebert of Lexington Books and the Rowman & Littlefield Publishing Group for their patience, assistance, and guidance through this book's publishing process.

For Michael Hewson

I have the enormous privilege of exploring the world with an ever-patient partner. To Ann Hewson—with huge thanks. I acknowledge a debt to the staff of the Humanities discipline in the CQ University, College of the Arts, who helped transition an Environmental Geography skillset into a much more rounded Environmental Humanities mind-set—and for clarity that the Earth needs the arts to communicate environmental policy—Amy Johnson, Benjamin Jones, Celeste Lawson, Jiaping Wu, Leanne Dodd, Lincoln Bertoli, Mike Danaher, Nicole Anae, Patrick Connor, and Stephen Butler.

Introduction

If the *superpower* of humanities is *storytelling*, a convenient question is this: in the information age, how useful has geographical media and digital mapping technology been to the *story* per se? Indeed, we are familiar with the idea that maps help express and explore *place*. And it is well known the human connection to place is foundational to well-being—particularly natural places. Further, sociological and, for First Nations people, physiological connection to *place* defines who we are. Place is collectively and all at once: moral fiber, ancestry, history, language, art, character, and culture. Accordingly, the discipline domains of humanities scholarship necessarily contain an undercurrent of *location*.

Here in this present book, we ask the reader to consider how geomedia and geographic information systems (GIS) advance the task of environmental humanities to educate, debate, and sway society in understanding, owning, and addressing the ecological emergency and environmental justice.

Deborah Bird Rose and colleagues (2012) felt that the role of environmental humanities was to contend with the growing awareness of the interconnectedness of social and ecological challenges. They argue that the environmental humanities drew on the humanities and social science enquiry tools (such as qualitative analysis) to unpack the underlying questions of meaning, value, responsibility, and purpose in times of changes such as that experienced in the present environmental emergency.

Fast et al. (2019, 89) define *geomedia* as an advanced set of digital media devices, hardware, and software that register and respond to a person's location, which has social and cultural ramifications. We use the word *geomedia* as a broader umbrella term for communication and entertainment technologies that utilize geographical data to educate, engage, and enable critical reflection. Geomedia then includes virtual globes and geotagged videos and images. These geographical data visualizations and evaluation extensions help connect humans with a complex repository of spatially enabled knowledge. For the humanities, geomedia is emerging as a crucial enabler of wicked problem-solving education.

And among all this geomedia, what *is* this thing called a GIS? Of course, the words give it away: *geographic*, or Earth's physical feature quantities; and *information system*—a repository of relational data. GIS is then a tin can of relatable things on and of the Earth. From a technology perspective, the now ubiquitous GIS allows us to visualize, analyze, and, indeed, create new knowledge with location-dependent data.

How has GIS helped the humanities in its task to synthesize cultural knowledge and be a vehicle for society's critical thinking? How does the science (and art) of digital mapping, and a spatial dimension, help develop society's narratives? Does it matter? Here we argue yes, yes—and yes. You will want to know the evidence, of course.

In recent decades *ecocriticism* has emerged from literary and cultural studies to cement a spot in the humanities toolkit. Garrard (2012) defines ecocriticism as a scholarship that explores the various ways in which the relationship between humans and the environment are both imagined and portrayed in the many media of cultural output. Ecocriticism unpacks the messages of ecology within artistic, literary, and other cultural productions of the past. Here, we explore the concept that geomedia is a powerful tool for ecocriticism based on a proven capacity to engage an audience. Indeed, authors in this book show that geomedia enhances a successful, society-shifting storytelling pedagogy.

GIS and geomedia are novel expressions for studying and analyzing global issues with an environmental humanities sensibility. As a means of curating data in multimedia forms, GIS provides the scholar with a worldwide representation of this information in ways that afford geo-relational study. By comparing geographic data with locative media tools, new data emerges, which sheds new light on environmental issues common to experiences in different parts of the world.

The discovery of relationships between historical events, human behavior, climate shifts, social, economic, and cultural adaptations and evolutions makes GIS a valuable geospatial study tool in the environmental humanities. The digital era of big data, map animations, multimedia, weblinks, and artifact archiving can all be explicitly situated to their points of origin or areas of study to provide an enhanced perspective as well as a new framework for environmental investigations.

The explorations and experiential observations of this book are part of a continuum that began when measuring the Earth became a "thing." In around 240 BCE, a librarian named Eratosthenes estimated what Aristotle had already thought: the Earth was round. Eratosthenes calculated the circumference of the Earth by comparing noontime sun shadows in deep water wells separated by a known distance. Since then, advances in measuring the Earth, and quantifying the anthropogenic impact on that Earth, have steadily progressed to the stage where now, geomedia does the heavy lifting of Earth-story communication. Indeed, the reader will recognize the importance of geomedia and location services to their ordinary and everyday life. Whether its consulting Google Maps, ordering an Uber, watching a Netflix movie, seeking Garmin directions or health advice, or feverishly watching your Amazon package tracking—to coin a real estate agent's phrase:—*it's all about location*. But, wait, wait—Netflix? Oh yes, Netflix needs to know your country of location to enforce the proper video license arrangements. All of these are GIS- or GPS-enabled products and services. All are becoming ubiquitous.

GIS can record things that happen—because things happen somewhere, and sometimes, to someone (Longley et al. 2015). The where, who, and when are core ingredients of GIS. However, underneath the GIS hood is an astonishing array of tools, techniques, and technology. The authors here take all that engine and suspension for granted and move on to exploring the efficacy of GIS for helping the humanities task of enquiry.

The chapters of the present book are arranged around three meta-themes. Looking through a spatial lens, closer chapters are related to each other more than chapters further away. The sharp readers among you will recognize here a tweaking of the purported *first law of geography* (Tobler 1970).

The co-editor and author of chapter 1 sets the tone for the central thesis for the book—that GIS and geomedia converge the manifold wisdom and knowledge from many stakeholders toward that ephemeral quality we in the humanities call *critical thinking*—in short, helping society make sense of complexity. He sets the scene for the book in describing GIS and a technological journey from spatial data visualization to undergirding an environmental humanities creative impact on shifting society to collective problem-solving.

The first meta-theme for this book, at the intersection of spatial technologies and environmental humanities, starts where it should—with the exploration by First Nation scholars exploring culture with geographical information methods. In chapters 2, 3, and 4, the question is posed: is GIS capable of visualizing the deep-seated connection to Country (and the big "C" here is essential, as is the lack of *the* before Country). We acknowledge, indeed embrace, the millennially long and deep cultural experience of First Nations people. In doing so, the authors ask if GIS advances the colonizing descendants' understanding of the visceral experience of the ancestral dislocation from Country? Does geomedia help the communities connect?

In chapter 2, a contributor and colleagues leave no stone unturned in exercising one of the essential tasks for postinvasion reconciliation—the need for *truth-telling*. Here, the reader hears how GIS is used to explore a history of colonizing invasion of a southern portion of *Terra Australis* in the late eighteenth century, usurping the original custodians, who had nurtured the landscape for at least 60,000 years.

The author of the third chapter journeys with First Nations peoples' community planning using GIS. The GIS technical capability of recording community assets is taken to another level to apply community concerns on how infrastructure can be utilized to address community problems and progresses sustainable community design by peoples used to practicing ecological kinship. Here the contributor introduces us to the idea that GIS data collection, information analysis, and knowledge synthesis can be participatory experiences leveraging collective wisdom. And on that journey, technological limitations and accessibility barriers are dealt with—after all, nothing is perfect.

A contributor in chapter 4 grapples with the use of GIS by Indigenous peoples to chart culture, claims, and Country. Here competing philosophies of land ownership and economics across cultural lines challenge the design schema for GIS and the subtle ways in which analysts consider spatial information. Even so, the connecting of Indigenous and non-Indigenous stakeholders in constructing location-tagged knowledge is paramount to the practical reconciling discussion and exploring the rights to land and livelihood. Technology has human users, and here, GIS is a cultural communication bridge and an enabler for social change. The author expands on the limitations of technology and helps many who will follow leverage the knowledge gained.

The second meta-theme inherent in the following three chapters (5, 6, and 7) revolve around the claims that GIS can be a humanities discipline line of enquiry—that digital maps can contain, indeed enable, an overlay of culture, justice, and moral choice in the exploration of lithospheric bound data elements.

In chapter 5, the author walks through a case study of water management spatial data in India and, with an affirming ecofeminist lens, finds that a lack of gender inclusivity can be overcome with some careful visualization assumptions. We discover that the western scientific data framework for GIS design can be enhanced by considering how space and data can be reimagined with feminist empowerment and how maps can achieve equity in the community use of critical natural resources. The key in this chapter is not so much what GIS can *do* as a technology but how agencies can *use* the GIS as praxis.

A contributor in chapter 6 examines how GIS provides a platform to hold, in happy tension, the emerging interdisciplinary scholarship of environmental humanities (EH). He expounds on the idea that an EH GIS is a methodological tool knitting ecocriticism, environmental history, and the philosophy of environmental ethics. In doing so the author contends with the need to break out of information system temporal and spatial strictures, and embrace an idea that GIS has a theoretical role for pursuing ecological justice and sustainability. The participatory GIS movement and the self-maintaining capability of complex information systems are employed to illustrate the *common good* for GIS. The map and GIS are repackaged and defined in an environmental humanities frame and acknowledges that geomedia is an evolution of education—but with impact. And *impact* is something that cultural practice has aimed for—sometimes with extraordinary success. In the 1980s, for example, just one seminal photograph of a wild river published in regional newspapers shaped environmental policy in Australia for the decades that followed (Dombrovskis 1979). The author reminds us that a humanities treatment of geomedia contributes to a more connected science communication toolkit.

And that's where chapter 7 by the co-editor takes up the reins. Can GIS be a communication tool? Science writers can indeed write, but can they cut through the polarization of society like cartoonists, artists, and creative writers can? Could maps tell stories? The author compares the "professional" standards-infused creation of GIS output with the craft of creative nonfiction and finds that the elements of craft overlap and complement. The mapmaker does well to study how society tells stories to itself, which might result in mapmaking that are an engaging and entertaining means of narrative.

The third meta-theme, chapters 8, 9, and 10, explore pedagogies for using GIS and geomedia in the humanities classroom (among other places, arguably any space where critical thinking needs enhancement). The challenge for teachers is to use geomedia and digital maps as a storyboard, as a movie screen, as a message mechanism to engender enthusiasm and connection with society. Here the *how* is explored with the *why*.

The contributor of chapter 8 explores the role of ecocinema and digital maps in encouraging a reflective culture that questions the status quo. His focus is on enabling student skills and competencies concerning sustainability. Sustainability is not always what pundits think it is—"sustainability" demands of society a wholistic rethink of natural resource use policy—and a realignment of a suite of values. The United Nations Sustainable Development Goals is the backdrop to the canvasing of the role of educators in sustainability. Educators influence the organic creation of change in the hearts and minds of their students, and this role of *eldership* involves story. Thus, the *change* role of story has always "been"—in any age, in any culture, in any medium, in these days arguably, video is the medium *du jour*. The author explores ecocinema teacher competencies via the theoretical basis and case studies.

Chapter 9 goes in an allied direction to chapter 8 and explores what it takes to consider an engaging digital map pedagogy in a health education setting. The contributor explores the experiences of the Planetary Health Film Lab that took place at the Dahdaleh Institute for Global Health Research, York

University, in Toronto, Canada. While the case study reflects much insight for the learning design of GIS-enabled climate change impact on health, the take-home talking points have a wide application in youth and adult education. Student participation emerges again—as does the role of place in society—and we are reminded that geomedia can enable community cohesion (as long as the technology limitations and cultural assumptions are worked through, as the author of chapter 5 reminds us). Here, the student's voice frames the outcomes achieved, thus indicating that technology outputs can be engaging and enabling—a recurring theme for a sound pedagogical reason.

Chapter 10 returns to the essential thesis, that GIS and geomedia have an emerging and central role in creating student engagement and delivering on the tropes of a humanities approach to exploring the craft of culture. The case study is a carefully planned research project into implementing geomedia to enhance the student acquisition of ecocriticism skills. The authors use the same eco-literature but new analytical tools and discuss the impact on the student experience in a two-year study they conducted with their undergraduate students at Toronto's York University.

In this present book, the authors impart their experience of using geographical technologies to advance the environmental humanities. We explore how geomedia and GIS contribute to unpacking underlying questions of meaning, value, responsibility, and purpose in times of changes experienced in the present environmental emergency. We examine the educational value chain of visualizing geographical information. We review how digital maps help pursue critical thinking and reflective culture. And last, we envisage an emerging nexus of spatial science and environmental humanities. Like any technology, the "fit for purpose" test is not just measuring the mastery of geomedia or mapping production skill but in its capacity to aid an outcome of community solving its many dilemmas. And like much technology, geomedia is just a tool and with the right mind and skilled hands, such tools evolve stories that shift society.

REFERENCES

Dombrovskis, Peter. 1979. "Morning Mist Rock Island Bend, Franklin River, Tasmania." Trove. Accessed 28/02/2018. http://nla.gov.au/nla.obj-147241885.

Fast, Karin, Emilia Ljungberg, and Lotta Braunerhielm. 2019. "On the social construction of geomedia technologies." *Communication and the Public* 4 (2): 89–99. https://doi.org/https://doi.org/10.1177/2057047319853049

Garrard, Greg. 2012. *Ecocriticism*. Edited by John Drakakis. *The New Critical Idiom*. Abingdon and New York: Routledge.

Longley, Paul A., Michael F. Goodchild, David J. Maguire, and David W. Rhind. 2015. *Geographic Information Science and Systems*. Wiley.

Lovelock, James. 1979. *Gaia: A New Look at Life on Earth*. Oxford University Press.

Rose, Deborah Bird, Thom van Dooren, Matthew Chrulew, Stuart Cooke, Matthew Kearnes, and Emily O'Gormand. 2012. "Thinking Through the Environment, Unsettling the Humanities." *Environmental Humanities* 1 (1): 1–5.

Tobler, Waldo. 1970. "A computer movie simulating urban growth in the Detroit region." *Economic Geography* 46 (2): 234–240.

Chapter One

GIS and the Environmental Humanities

How Citizen Scientists, Civil Servants, and Researchers Are Teaming Up to Study and Solve Environmental Issues

Mark Terry

Ever since Roger Tomlinson conceived the geographic information system (GIS) as a computerized method and technique for storing and interpreting mapped data for a project known as the Canada Land Inventory in 1962 (Tomlinson 1967), a vast array of academics has been harnessing the power of GIS in the production of knowledge.

In a documentary film produced by the National Film Board of Canada called *Data for Decision* (Millar 1968), Tomlinson explains the concept of the GIS as a means of understanding vast volumes of data with the assistance of the computer, equating it to a new farmer trying to understanding their immense property:

"He really doesn't know how big the farm is, perhaps he doesn't know much about the soil. He doesn't have too much of an idea about the climate and he's not really sure if there's usable water. And yet, he has to make decisions that will let him plant the right seed and grow enough food to support his family. And if you think that would be a problem on an ordinary sized farm, think what it would be like with a million square miles . . . We like to think we have a system that can accept the information, can store it, can analyze it and present the results in a usable form. A system that can do this not in years, but in hours" (Millar 1968).

It is noteworthy that this first public manifestation of GIS technologies was for a government environmental project, an assessment of the land of Canada to assist policymakers in servicing the needs of the country and its people. Even the farming example speaks to the use of GIS in an environmental scenario assisting the farmer in decision-making. As well, the title of the film suggests that GIS should be used by governments to aid in their decision-making process (i.e., policy creation). This documentary film not only introduces the world to GIS but also suggests how it is to be used.

Advancements in computer technology since then have expanded the affordances of GIS to allow for even greater speed, representation, and analysis of data, not exclusively related to geography. Today we have come full circle from this original intent as one field of research and scholarship in particular seems well-suited to GIS platforms—the environmental humanities. Earth scientists, climatologists, meteorologists, and similar planetary health researchers—hereinafter referred to collectively as "environmentologists"—can now use tools developed in the environmental humanities to create cartographic databases of information comprising statistics, texts, and media such as photography and film to better understand the research, impacts, and causes of environmental phenomenon on a global basis. This platform synthesizes the data in unique ways that reveal explicit trends as well as implicit revelations of new data that emerge from relating temporal and spatial information afforded by GIS metadata.

Originally, GIS was used as a way of mapping and understanding geographic features, then seen as a tool for historians as dated data geolocated in similar areas would reveal changes over time. These changes could then be interpreted to determine the causes of these changes or at least point the researcher in a direction to look in more focused disciplines of study to uncover the causes of these changes. A specific glacier, for example, examined over a traditional dataset of thirty years, would yield an annual database of measurements related to decline. The critical historian could then use this data to establish a rate of decline over this period which would then, in turn, provide a reasonable forecast of future decline. This becomes important when projecting the negative impacts this decline has on ecosystems and neighboring communities. Knowing this may help governments prepare for such eventualities and take precautions that protect the environments of flora, fauna, and people.

Soon, geographers began to look toward the spatial relations between points on the map more closely. Data embedded in these pins can be compared to and contrasted with data in other locations providing the environmentologist with another investigative tool to procure new data and to identify new influencers on empirical observations. By combining time and space this way larger patterns emerge in this cumulative database that services both the initial data collector and the subsequent analyst.

While the dimensions of time and space are cornerstones of GIS architecture, many environmental humanists add a third dimension: attribute. Once an environmental feature is identified by time and place, a description of this feature is provided to assist the analyst in understanding the changes to the feature in the place of occurrence and at the time of occurrence. With multiple entries of attributes in these GIS layers or metadata, a reliable dataset can reveal new data and yield a fuller comprehension of the profiled environmental phenomenon.

This works fine for easily observable and measurable geographic features like terrain, mountains, and bodies of water, but for more abstract data related to environmental elements such as climate, air quality, weather, and temperature, data recorded independent of the GIS process can be assimilated in the map though this interdisciplinary process of data collection, and this is where things become interesting. A virtually unlimited amount of data in almost any digital and multimedia format can be added as layers, sometimes even live video feeds to show the analyst what is happening at the moment of study.

The environmentologist uses both history (time) and geography (space) to contextualize each other. By relating map units spatially, changes in the environment of each place (climate, weather, terrain, population distributions and relocations) can provide sources of forces that act upon the data collected over time. Conversely, reasons for changes in statistical data within a temporal dataset can be attributed to specific anomalies assigned to each space. If we return to our glacier example, a rate of decline in the Arctic of 10 percent per year related to a similar glacier in Antarctica that melts at a rate of 20 percent per year yields the obvious conclusion that glaciers in Antarctica are melting twice as fast as those in the Arctic. What this relational process does not address is why; however, when we apply the environmental conditions (historical weather patterns, temperature records, meteorological events, etc.) of each place, respective impact factors can be assigned and an additional level of data is achieved providing the scientist—and the policymaker—with a fuller understanding of an ecological issue that is at once regional, but also attributable globally.

When this kind of data is added to the equation, multimedia components such as film, photography, and live feeds can provide a visual context to the statistical data, essentially putting a face to the name.

Ultimately, environmentology serves to identify areas of concern with a goal of providing the necessary data to inform and influence progressive legislative change to improve these areas and the lives of all those impacted by the issue directly and indirectly. Polluted water upstream can poison the drinking water for communities further downstream, even thousands of miles away. Those at the source may live in a relatively pollution-free environment, but the fluid ecosystem of shared water and air can adversely affect the health of those elsewhere and, in fact, the health of the planet as a whole.

With such high stakes, we now see the environmental humanities claim its place in GIS databases in a variety of disciplines all contributing to provide the policymaker with abundant information intended to translate scholarly research into comprehensible knowledge to be used to empower the legislator with an understanding that may not be otherwise achieved by those without formal academic and scientific training.

This chapter explores the relationship between the environmental humanist as map designer and curator, the environmental scientist as data collector, and the environmental policymaker as analyst. All three partners collaborate in a new process of observation, reporting, and communication that demystifies global environmental issues in a journalist-teacher hybrid organization designed to represent researched phenomenon, educate the user, and ultimately inform and influence those charged with creating essential and progressive environmental policy.

In particular, we will examine the role of the "citizen scientist" as data collector and their support of environmental science research in three case studies that aim to democratize and globalize the process of environmental research among the policymaker, the scientist, and the global citizen.

This triumvirate of actors and their mutually interactive behaviors will be examined in three case studies: the Ecological Footprint Initiative, the eBird digital database, and the Youth Climate Report.

PROJECT: THE ECOLOGICAL FOOTPRINT INITIATIVE

GIS Platform: Google My Maps

An ambitious research project out of Toronto's York University seeks to measure how humanity's consumption of renewable resources has changed since the first Earth Day in 1970.

The project, known as the Ecological Footprint Initiative (EFI), involves faculty, students, and staff from the Faculty of Environmental and Urban Change in reviewing and codifying more than a thousand scientific papers related to a measurement for "the *demand* on and *supply* of nature" (*Global Footprint Network* 2021). The metric was conceived by Mathias Wackernagel and William Rees at the University of British Columbia in 1990 and has been the basis of hundreds of research papers on environmental consumption. It is widely used by scientists, businesses, governments, individuals, and institutions working to monitor ecological resource use and advance sustainable development. The York University project narrowed its field of analysis to the top 200 most-cited papers in the batch representing research from 1992 to 2019.

To calculate the demand of an area by those who live there, an ecological footprint (EF) is used to measure the "ecological assets that a given population requires to produce the natural resources it consumes (including plant-based food and fiber products, livestock and fish products, timber and other forest products, space for urban infrastructure) and to absorb its waste, especially carbon emissions" (*Global Footprint Network* 2021). This calculation involves the tracking of six key categories: cropland, grazing land, fishing grounds, built-up land, forest area, and carbon demand on land. To calculate the available supply, an area's biocapacity (BC) is assessed based on "the productivity of its ecological assets (including cropland, grazing land, forest land, fishing grounds, and built-up land)" (*Global Footprint Network* 2021).

This approach to assessing supply and demand became the basis for the York University EFI project when it established coding forms for participating researchers to use in reviewing the scientific papers it sampled over a twenty-seven-year period since the ecological footprint metric was introduced. It is important to note here that most of the researchers analyzing these papers were graduate students, emerging scientists in their own right, but currently serving as citizen scientists at this stage in their academic career.

The methodology of review involves teams of two to three researchers reading a selection of papers weekly over a period of two years (2018 to 2020). At the end of each week, the team members met to compare their respective codifying forms to come to a consensus. The categories used to codify the papers include:

- Identifying the paper's purpose as it relates to the ecological footprint (EF) and biocapacity data (BC) as a critique, response to a critique, or an innovation or update to an aspect of the ecological footprint metric.
- How the paper uses the metric to inform a specific decision/plan/policy; to compare/use BC/EF with other sustainability measures; and to use EF/BC in communications and/or engagement/deliberations.
- What is the paper's scope and scale of use of EF and BC with respect to individuals or households or families or socioeconomic segments; non-profit institution(s) such as university, hospital, government operations; and business/industry/enterprise including entire sector?
- What is the geographic coverage of the paper's research: city or town, state, province or subnational region, national, regional, and global?
- Which EF components the paper consider: cropland, grazing land, fishing grounds, built-up land, forest area, carbon footprint (demand on land), or all?

- What is the methodology of the study: EF calculated using global footprint network (GFN) methodology or other?
- Which data was used: EF data derived from the GFN project or other?
- Does the paper consider biocapacity (BC) using the same EF components and GFN methodology and data or others?
- Concluding recommendations:
  - Further research to extend/advance EF/BC accounting/applications
  - Use of EF/BC by researchers/governments/businesses/individuals
  - Less use of EF/BC accounting/applications

By using a GIS platform for much of this data, spatial and temporal perspectives are available to provide an advanced method of analysis that promise to yield new data when the pins are related to each other. As a digital resource, each pin on the EFI map project includes the original paper reviewed by the EFI team at York University and, when possible, multimedia components (photography and film). The metadata for each pin represents the data derived from the consensus reports, the content of which reflects many of the categories listed above (see figure 1.1).

The EFI project examines research from eighty different countries and territories and identifies twenty-four multidisciplinary subject areas shown here in table 1.1.

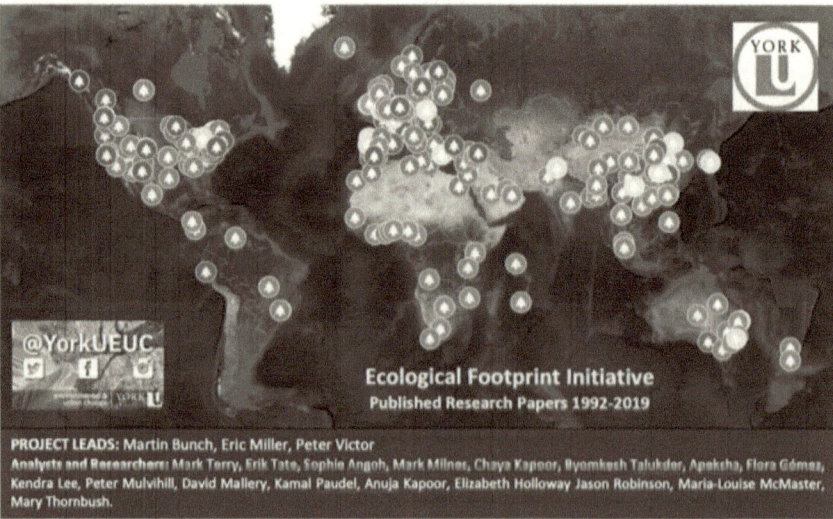

Figure 1.1. GIS map of more than 400 entries representing 200 scientific research papers related to the Ecological Footprint method used to measure human demand on natural capital. The colored pins represent the areas covered in each paper's respective research.

Table 1.1: Number of references per subject area (*List* 2020)

Subject Area	Number of References
Agricultural and Biological Sciences	221
Arts and Humanities	16
Biochemistry, Genetics, and Molecular Biology	18
Business, Management, and Accounting	103
Chemical Engineering	13
Chemistry	10
Computer Science	19
Decision Sciences	109
Earth and Planetary Sciences	53
Economics, Econometrics, and Finance	162
Energy	187
Engineering	148
Environmental Science	820
Immunology and Microbiology	5
Materials Science	7
Mathematics	12
Medicine	17
Multidisciplinary	7
Neuroscience	4
Nursing	4
Pharmacology, Toxicology, and Pharmaceutics	8
Physics and Astronomy	3
Psychology	11
Social Sciences	420

In a 2018 press release issued by York University, EF creator Mathias Wackernagel praised the decision to headquarter the EFI at York for its ability to create ways of making the data more accessible:

By establishing the National Footprint Accounts at York University—a vibrant, independent, well-governed, and respected academic leader in sustainability—the accounts will become even more trusted and effective. After spreading the use and recognition of ecological footprint accounting around the globe, this new placement at York will turn these accounts into an even more incontrovertible reference for public and private decision-making in support of sustainability. (Media Relations 2018).

A GIS map of the research and its findings is one way of doing just that. In considering the end user of this data—the policymaker—an open source GIS platform is necessary to afford access to everyone in the global community policy at all levels of government and policy advisory. Google My Maps, while not open source, per se, can provide subscription-free access to users and supports a robust searchability function many policymakers who are not familiar with scientific jargon have identified as being essential for their work (Terry 2020, 178).

The EFI GIS project was released in the spring of 2021. As defined earlier in this chapter, this project is intended to serve as a communications tool to bridge the gap between science and policy in the extensively researched and legislated fields of carbon, ecological, and global environmental footprints left by humans and the supply and demand relationship between humans and their planet. By involving students as citizen scientists in the review and coding process, the triumvirate of environmentology partners—the humanist (data curator and map designer), the scientist (paper authors and student researchers), and policymaker (research papers and project funders, and, ultimately, end-user of the project's data)—come together as a team to collect, analyze, and communicate findings related to a global database of environmental research related to measurements of ecological footprints worldwide.

PROJECT: EBIRD

GIS Platform: Google Maps

On the conservation side of the environmental humanities, a GIS project launched in 2002 by the Cornell Lab of Ornithology at Cornell University and the National Audubon Society called eBird is a crowd-sourced database using the global community of bird-watchers as citizen scientists. Contributions made by amateur and professional bird-watchers, known to themselves as birders, are done in real time around the world in a democratizing system that supports scientific research in migration patterns, population numbers, and location identification.

GIS maps created by eBird locate thousands of species of bird worldwide and offer a search option to filter markers by specific species names. Other filters allow you to see where the species has been spotted by time: year-round (current year, all years, or past decades). Specific months and years can also be selected to provide an historical perspective to breeding areas and migration patterns. The maps also offer location filters where you

GIS and the Environmental Humanities 17

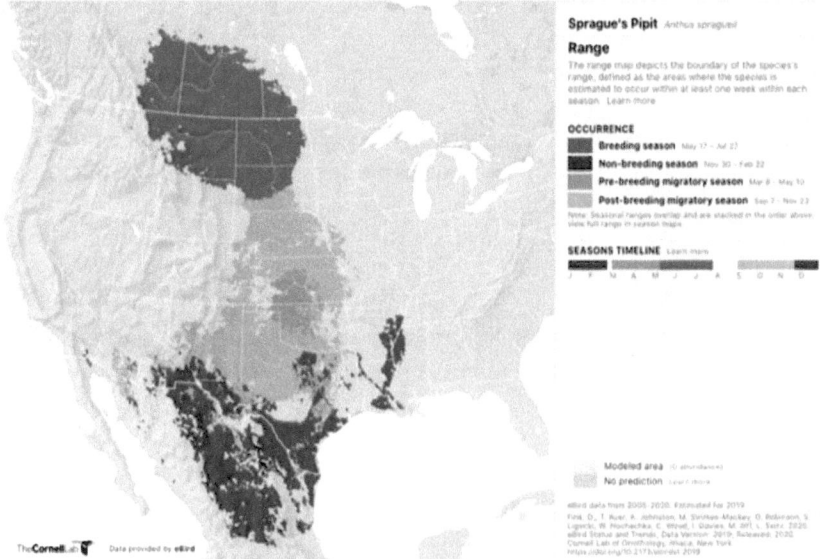

Figure 1.2. Example of an eBird GIS map showing the range of distribution of a species known as the Sprague's Pipit in North America during breeding and non-breeding seasons as well as pre- and post-breeding migratory seasons.

can enter a continent, country, state, or province, even a city to show the population numbers of any species at any time (see figure 1.2).

Citizen science projects are not the same as academic science projects carried out by professionals with specific skills and credentials in their respective fields of study. As a result, citizen science projects—those performed by amateur scientists, usually termed "enthusiasts"—can often fail due to flaws in methodologies, observation recordings, and ill-informed hypotheses and conclusions. With these shortcomings in mind, the creators of eBird expanded on the management of the citizen scientist observations to control and transform the crowd-sourced data in a more traditional and scientifically reliable manner. The resulting GIS map reflects these refinements through a myriad of layers identified in collaboration with stakeholders: participants who collect data, researchers analyzing the data, agencies adopting policies based on the data, and land managers taking direct conservation action (Sullivan et al. 2014). The specific collaboration results in data that are accessible to academic and nonacademic audiences alike and the GIS platform visualizes the data, with layers democratized for all, effectively bridging the frequently identified communications gap between science and policy.

Instead of trying to make scientists out of nonscientists, eBird took the community approach involving the active participation of all stakeholders, including the citizen scientist whose field work is crucial to the database of a project designed to account for all the birds of the world, their locations, and migratory patterns. This participatory perspective revealed a new methodology of data collection and management to the scientist:

Ultimately, these collaborations have enabled us to increase both the quantity and the quality of useful data available for analysis. Further, this broad spectrum of intellectual contributions and applications has fundamentally changed our view of the project, which we now see as a collective enterprise. (Sullivan et al. 2014, 31)

While this collaboration is enthusiastically pursued by the scientist and the volunteer, concern over the reliability is not without address. An article in the *Journal of Field Ornithology* reports that since amateurs are being involved in a scientific process, safeguard measures exist to ensure the accuracy of the data collected:

To further improve data quality, eBird enlists the help of volunteer experts who develop regional filters based on the chosen spatiotemporal coordinates and date of observations. These filters set limits for the species observed as well as a maximum number of individuals for that species for those submitting volunteer observations. If a count surpasses the max, or a bird is reported outside the specified date range, the record is flagged for further scrutiny by an expert reviewer. (Callaghan, Gawlik, 2015, 299).

In any GIS project, the qualitative and quantitative natures of the data are crucial. Critics caution GIS curators to be wary of falling into the trap of providing mere "collection over selection" (Druick 2018, 401), and cognizant of this pitfall, those behind eBird have incorporated the failsafe precaution of fact-checking the amateurs with professionals. In a 2015 study of the eBird GIS project conducted by Corey T. Callaghan and Dale E. Gawlik titled "Efficacy of eBird Data as an Aid in Conservation Planning and Monitoring" for the *Journal of Field Ornithology*, the researchers found that the controls put in place ensured that the data was just as reliable as those surveys of birds done exclusively by academics:

Our study demonstrated that eBird is not just applicable to questions at a large scale, but can also be applied to questions about small-scale individual projects. When controlling for observer effort, our estimates of diversity from eBird did not differ from those based on standardized surveys. (Callaghan and Gawlik 2015, 303).

As an added measure, the scientists behind eBird offer interactive tools that provide methodological and epistemological direction in the crowd-sourced process of observation and record:

> Our research and analysis teams suggest best practices for data collection, and our data quality process helps participants realize when they've made a mistake,

helping them become better birders. Our web tools are geared toward providing user reward. All these things lead to increased engagement and data volume, which in turn creates better data products and more informed conservation outcomes. (Callaghan, Gawlik 2015, 303)

Collaboration with policy stakeholders reveals a reciprocal engagement in which data is both received and used by government agencies as well as provided by the agencies to the GIS map directly. One such agency, the National Aeronautics and Space Administration (NASA) in the United States has a formal partnership with eBird in which they provide space-based data such as land cover, water cover, elevation, and topography to the map enhancing both the spatial and temporal analysis affordances of the observational data of the global bird populations by both the scientist and the citizen-scientist (NASA 2021). The additional data serves both the scientific community as well as the environmental policymaker better than other surveys without this collaborative research structure.

Daniel Fink, a member of the eBird team and senior research associate and statistician at Cornell University, believes this partnership is "critical" in providing those who require accurate data for the creation of informed policy. "We use land cover information to produce maps that predict where blue jays will occur across the landscape in places where people haven't observed them directly. When we look at these maps, the continuous coverage they provide is there because of the continuity of the NASA data products" (Cole 2018, 1).

Another government partner in the eBird collective are the Canadian Arctic regions of the Northwest Territories and Nunavut through their Bird Checklist Survey, a research project launched in 1995 by the Canadian Wildlife Service in response to needs identified in the Canadian Landbird Monitoring Strategy for better monitoring of Arctic breeding birds and birds with primarily northern ranges (eBird 2021). The survey joined the eBird project in 2012 to share data mutually and better represent this data with the GIS technologies eBird has initiated and refined since its launch in 2002.

Returning to the country of origin for GIS technology, this government program is similar in scope to the first GIS project, the Canada Land Inventory, in that Bird Checklist Survey is a nationwide accounting of birds in Canada's Arctic territories. All historical data collected by the government program was added to the eBird database along with current data and an invitation to the citizen-scientist to contribute more going forward. A rich database of bird population in a very specific place not commonly surveyed with a temporal dataset of twenty-five years is now available for policy and public access within the rich search option of the eBird GIS platform.

To assist the citizen-scientist, the Canada Wildlife Service makes available a checklist comprising of all known species with spaces for new ones. The form asks the volunteer researcher for the all-important "where and when" questions requesting longitude and latitude coordinates to identify place and the date (month/day/year) and time (hour of the day and the duration of observation in hours and minutes) to identify time. These two dimensions are key to any GIS project and are used here to reveal historical trends in Arctic places that inform projections of future population densities, nesting areas, and migration patterns and routes.

Government officials in Canada benefit directly from the synthesis of this data made within the eBird GIS interface affording a more accurate survey of population distributions not just in the Canadian Arctic, but elsewhere in the country. This assists the environmental policymaker in identifying migratory patterns outside of the Northwest Territories and Nunavut as they occur throughout the rest of Canada.

PROJECT: THE YOUTH CLIMATE REPORT

GIS Platform: Google My Maps

Another example of crowd-sourced projects made in collaboration with the policymaker is the Youth Climate Report, a database of documentary films made by the global community of youth for the United Nations. While this project has been serving the UN climate summits since 2015, its use of the citizen-scientist—in this case, more of a citizen journalist—is very similar to the eBird approach to training its global network of contributors.

According to Scott Mackenzie and Anna Westerstahl Stenport, digital media projects such as this "take on a larger role in climate communication" (Mackenzie and Stenport 2020, 90). They go on to describe the Youth Climate Report as a "transnational, participatory media, interactive digital project" structured to present "a series of youth reports from around the world about climate change, which includes the changes that the Arctic is undergoing, featuring, collectively, a network of globally linked, interactive videos using (GIS technologies)" (Mackenzie and Stenport 2020, 90).

Working with an age bracket of eighteen to thirty, the United Nations Framework Convention on Climate Change (UNFCCC) office issues a call for entries to its annual Global Youth Video Competition. The contest is called Earthbeat and is organized by the youth office of the UNFCCC and youth4planet [sic], a not-for-profit organization based in Luxembourg that offers "a framework and a platform engaging them in the complex process of cooperative filmmaking—in schools, colleges, workplaces (and) youth organisations" (youth4planet 2021).

The annual competition presents two to three themes, usually related to the Sustainable Development Goals, for student filmmakers to follow. Hundreds of films are entered and the top twenty in each category are added to the YCR project. To support the novice filmmaker, the YCR website (youthclimatereport.org) offers tips in shooting, interviewing, and editing to ensure the content of the project has a relatively unified look. Once the films are uploaded to their respective longitude and latitude coordinates, metadata related to the content of each film is added to the project's pins. When available, links to scientific papers published by the interview subjects in the films are added together with links to their research institute, environmental organization, or university. The year of production and the country name are added for easy access to time and place for the policymaker looking for data related to specific locations and years (see figure 1.3).

In addition to the competition, other sources of curation for the YCR GIS project includes partnerships with the Foundation for Environmental Education and its Young Reporters for the Environment program, World News Day and its #NextGen Video Challenge, and Wilfrid Laurier University's Documentary Filmmaking for the United Nations program. Another program that involves training, workshops, and experiential education is the Planetary Health Film Lab, a research project at the Dahdaleh Institute for Global Health Research at York University in Toronto, Canada. This initiative brings together disparate groups of youth in an intensive, one-week workshop to learn the theory and practice behind producing documentary shorts for the

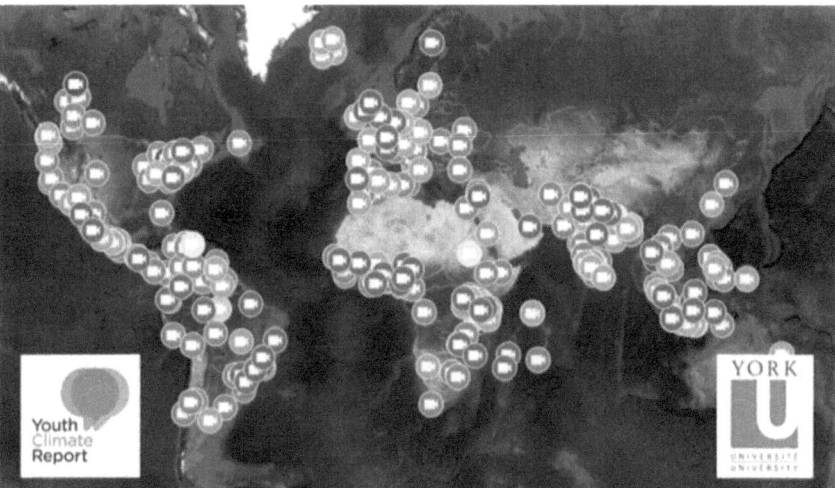

Figure 1.3. The Youth Climate Report GIS Project. Coloured pins represent the years of production of documentary short films produced by the global community of youth from 2008 to the present.

YCR project. Instead of adhering to a specific theme, participants are invited to tell stories related to environmental issues and planetary health from their home countries and communities. The first Planetary Health Film Lab took place in February 2020 with seven participants: two from Ecuador, and one each from Canada, Colombia, Italy, India, and Australia. In addition to receiving training on the mechanics of filmmaking, the participants also learn how to make a film specific to an audience of policymakers to understand how to deliver content that accelerates the path to progressive policy change.

This collaboration between the data collector (YCR), the citizen journalist (the participants), and the policymaker (the United Nations) establishes an ecology of communication and education which supports the respective goals of each actor. The YCR project is a remediation of the documentary film known as a Geo-Doc (Terry 2020, 3), a multilinear, interactive, database documentary film project presented on a platform of a geographic information system map of the world. The Geo-Doc concept uses the affordances of GIS technology to tell stories with additional dimensions, such as multilinear storytelling and implicit narratives, employing documentary film in the service of communications in an altogether new way. The United Nations seems to agree. The Youth Climate Report was awarded an honorable mention by the panel of international judges for the UN's Sustainable Development Goals Action Awards, the only GIS project to win the award (SDG Action Awards 2021, 3):

As new advancements in the field of geomedia expanded the multimedia components of GIS content, video became a valuable new element. This afforded the (YCR) project a platform that best suited the international policymaker. Unlike other database documentary projects that require exploration throughout the host website, the Geo-Doc format allows a bird's-eye view of the entire world and its corresponding film units. UN delegates could now easily select those film projects that related directly to the regions of the world that interested them most. (Terry 2020, 1)

Much like eBird, content is curated from global volunteers and organizers assist these young reporters with tips on its website and training workshops that reflect the collaborative nature of the overall project.

A typical film lab begins with an orientation day upon arrival by the participants. The next three mornings participants are taught by a series of scholars and professionals in the fields of communication, international policy, film production, digital media, and environmental studies. Days 4 and 5 are spent applying the knowledge they had been given in the production of their film reports. Professional support was provided by camera operators, film editors, documentary filmmakers, and sound recordists to assist the new filmmakers during this period of the film lab. By the end of the week, participants premiere and introduce their films to the public at a micro film festival. Shortly after, their films are uploaded to the UNFCCC's Youth Climate Report project adding their voices to the global community of youth.

Other film labs focus on amplifying Indigenous voices in circumpolar Arctic communities and in Ecuador. Young filmmakers from the circumpolar Arctic participated in the Film Lab to tell stories related to planetary health in their home communities and in the summer of 2022 the Indigenous communities in Ecuador will feature twenty young filmmakers telling their stories in their native languages of Kichwa and Shuar. As environmental issues in this part of the world are measurably different than the rest of the world, so too, are their impacts on the Indigenous people that live there. These stories become critical in the global conversation of planetary health and having these stories represented by film in a single digital space of a GIS mapping project affords the policymaker with the visible evidence necessary to assimilate the data in a more humane way. The interdisciplinary presentation of data in this regard provides a variety of perspectives and new data and metadata that contribute to a fuller understanding and can lead to better solutions that take into account the anthropogenic contributions to the ecosystem of the planet. As of June 2022, there are more than 620 videos and related metadata on the Youth Climate Report GIS map.

CONCLUSION

In all three case studies, the collaboration of three key partners is evident: the environmental humanist as map designer and curator, the environmental scientist as data collector, and the environmental policymaker as analyst. The first GIS project was a government project, but since then, the technology has become more accessible and intuitive and the contributors more democratized. The current triumvirate of GIS makers and users inform each other with data collection, data analysis, and, ultimately, data implementation in policy.

Unlike a traditional documentary film that intends to do the same thing by "speaking truth *to* power," the environmental humanities is recruiting the service of the changemaker in the process of production, in effect, now "speaking truth *with* power." Rather than an end user of content collected and presented on a GIS map by professional researchers alone, policymaking partners are becoming involved in the beginning as a project funder, content director, and trainer of citizen-scientist contributors.

With the various disciplines of this humanist approach, data takes on various forms—statistical, visual, textual, historical, temporal, spatial, process-relational—and with the global citizen, the scientific community, and the policymaker all contributing equally at all stages of the project, we see the emergence of a model of collaboration that is yielding accelerated paths to progressive environmental change worldwide.

The GIS projects examined in this chapter address specific questions, but when populated by data represented by the various disciplines of study of the environmental humanities, new questions emerge and may even spawn new maps in answering these new questions. Jim Herries, an applied geographer with ESRI, a GIS software developer, supports this multidisciplinary approach to map creation:

> We always say that a good map answers a specific question, but a great map inspires three more. Questioning the map is the key to discovery, and it happens so quickly and interactively these days. Someone can ask to see data in a new way or to adjust a key indicator and we can all see how the map reacts when anchored around a new goal. (Leadbeater 2019)

Seeing data in new ways is the objective of the environmental humanist when contributing to GIS projects. As evidenced by the collaboration of the triumvirate of environmentologists profiled in this chapter's case studies, this emerging role is developing new ways of researching, analyzing, and providing solutions to environmental issues that affect us all.

REFERENCES

Callaghan, C., and D. Gawlik. "Efficacy of eBird Data as an Aid in Conservation Planning and Monitoring." *Journal of Field Ornithology* 86, no. 4 (2015). https://doi.org/10.1111/jofo.12121.

"Canada, Environment and Climate Change." May 26, 2017. canada.ca. Accessed January 30, 2021. https://www.canada.ca/en/environment-climate-change/services/bird-surveys/landbird/ebird-northwest-territories-nunavut.html.

Cole, Steve. "North American Bird Maps Enter the Space Age." NASA (2018). NASA.gov, Accessed January 30, 2020. https://www.nasa.gov/feature/north-american-bird-maps-enter-the-space-age.

Druick, Zoë. "'A Wide-Angle View of Fragile Earth': Capitalist Aesthetics in the Work of Yann Arthus-Bertrand." *Open Cultural Studies* 2, no. 1 (2018): 396–405. Accessed February 7, 2021. https://doi.org/10.1515/culture-2018-0036.

eBird Species Map, 2002 to Present. Ithaca, NY: Cornell Lab of Ornithology.

Fink, D., T. Auer, A. Johnston, M. Strimas-Mackey, O. Robinson, S. Ligocki, W. Hochachka, C. Wood, I. Davies, M. Iliff, and L. Seitz. *Sprague's Pipit: Range*. The Cornell Lab of Ornithology, Ithaca, New York, 2020. Accessed January 30, 2021. https://doi.org/10.2173/ebirdst.2019.

Footprintnetwork.org. *Global Footprint Network.* (2022). Accessed January 30 2022. https://www.footprintnetwork.org.

"How the Footprint Works." Global Footprint Network (2022). Accessed January 30, 2022. https://www.footprintnetwork.org/our-work/ecological-footprint/?__hstc=207509324.38bf2a8d6b4848ea0a4553a6e221927e.1597989826356.1597989826356.1597989826356.1&__hssc=207509324.1.1597989826356&__hsfp=4040258624.

Leadbeater, Richard. "Maps Inform Public Policy, Help Turn Plans into Actions." *GIS ForRacial Equity* (2019). Accessed January 30, 2021. https://gis-for-racialequity.hub.arcgis.com/datasets/maps-inform-public-policy-help-turn-plans-into-actions.

List of Subject Areas and Number of References of Reviewed Papers. June 6, 2020. Toronto: York University,

Mackenzie, Scott, and Anna Westerstahl Stenport. "Visualizing Climate Change in the Arctic and Beyond: Participatory Media and the United Nations Conference of the Parties (COP), and Interactive Indigenous Arctic Media." *Journal of Environmental Media* 1, no. 1 (2020). Accessed January 31, 2021. https://doi.org/10.1386/jem_00007_1.

Media Relations. "York U to Calculate Progress on Key UN Sustainable Development Goals." April 17, 2018. Accessed: January 30, 2021. https://www.yorku.ca/media/2018/04/york-u-to-calculate-progress-on-key-un-sustainable-development-goals.

Millar, David. *Data for Decision*. National Film Board of Canada, 1968.

NASA. "North American Bird Map Enters the Space Age." December 19, 2018. Accessed January 31, 2021. https://www.nasa.gov/feature/north-american-bird-maps-enter-the-space-age.

SDG Action Awards. 2021. Honourable Mentions. Accessed January 20, 2021. http://sdgactionawards.org/youth-climate-report.

SDG Action Awards. "Youth Climate Report." (2021). Accessed February 7, 2021. https://sdgactionawards.org/youth-climate-report.

Speak Truth to Power. A Quaker Search for an Alternative to Violence. American Friends Service Committee 1955.

"Storytelling." youth4planet.com. Accessed December 2, 2021. https://youth4planet.com/storytelling.

Sullivan, Brian L., Christopher L. Wood, Marshall J. Iliff, Rick E. Bonney, Daniel Fink, and Steve Kelling. "The eBird Enterprise: An Integrated Approach to Development and Application of Citizen Science.. *Biological Conservation* 169 (2014): 31–40. Accessed January 31, 2021. doi:10.1016/j.biocon.2013.11.003.

Terry, Mark. "The Geo-Doc: A Locative Approach to Remediating the Genre." In *The Geo-Doc: Geomedia, Documentary Film, and Social Change*. Basingstoke: Palgrave Macmillan, 2020.

———. "Amplifying the Voice of Youth through Planetary Health Films." *The Lancet Planetary Health* 4, no. 12 (2020). Accessed February 7, 2021. https://doi.org/10.1016/S2542-5196(20)30248-5.

———. *The Youth Climate Report GIS Project*. Youth Climate Report, Toronto, Ontario, 2015. Accessed February 1, 2021. http://youthclimatereport.org.

———. *The Geo-Doc: Geomedia, Documentary Film, and Social Change*. Basingstoke: Palgrave Macmillan, 2020.

———. *EFI GIS Map.* Youth Climate Report. Toronto, Ontario, 2021. Accessed February 1, 2021. http://youthclimatereport.org/efi.

Tomlinson, Roger F. "An Introduction to the Geographic Information System of the Canada Land Inventory." In *Department of Forestry and Rural Development.* Ottawa, ON: Government of Canada, 1967.

Youth Climate Report. 2015 to Present. Bonn, Germany: United Nations Climate Change.

Chapter Two

Wadawurrung Dja

The Ethnography and Biogeography of Pre-Colonization Wadawurrung Country in a Digital Realm

Susan Ryan, David S. Jones, Murray Herron, and Phillip Roös

SURVEYING THE COUNTRY AND LANDSCAPE

Seeking Country within a multiplicity of layers of information is natural to Australian Aboriginals when explaining their Country. Country is not simply "landscape," nor land, nor seas, nor waters, nor blue skies, nor night skies, and so on, in the Western definitional abstraction of the word landscape (Jones et al. 2021, 11–18). Rather, it something more complex in its four-dimensional reality and composition (Rose 1996; Nicholson and Jones 2021, 508–525).

"Country is a place that is not easy to be defined as 'place' like in Western dictionaries, nor is encapsulated in typologies like Meinig's 10 ways of seeing or the Australia ICOMOS's *Burra Charter"* (Meinig 1979, 33–48; Australia ICOMOS 2013). The latter adeptly textually modified by Australia ICOMOS membership to better accommodate Aboriginal-rich "places," expressed "Place means a geographically defined area. It may include elements, objects, spaces, and views. Place may have tangible and intangible dimensions" (Australia ICOMOS 2013). This definition includes an "'Explanatory Note'" [whereby] Place has a broad scope and includes natural and cultural features. Place can be large or small: for example, a memorial, a tree, an individual building or group of buildings, the location of an historical event, an urban area or town, a cultural landscape, a garden, an industrial plant, a shipwreck, a site with *in situ* remains, a stone arrangement, a road or travel route, a community meeting place, a site with spiritual or religious connections" (Australia ICOMOS 2013; Jones et al. 2021, 11–18). These definitional differences are evolving in Australia but are still worlds apart.

The comprehension gulf is that each Aboriginal is resident inside their Country as a custodial participant waiting in anticipation of the return of their ancestors. Each serves as a surrogate equal holistic eco-system participant. This is markedly distinct from the landscape being subservient to human endeavors and being translated into a commodity. The former is what underpins the complexity in translating Aboriginal culture, knowledge systems, and lore and relationships, in Australia in times past to the Dreaming and into today and into the future (Rose 1996).

"Our words are clear in their logic here as they are nested in non-Western notions of ownership, land, truth, and time; the latter works in past present future but also in physical, spiritual and genetic time" (Nicholson and Jones 2019; Powell et al, 2019, 44–84). Additionally, recognize that "Country" is not one nation encompassing the Australian continent and seas, but rather there are over 250 Countries across Australia, each with their own language/dialect, protocols, stories, knowledge, "history," but they are all somewhat interconnected analogously like a tapestry bearing different fabric pieces and colored threads that are interwoven. Stress one thread and you stress the whole tapestry.

Recognizing this intellectual and philosophical context, a key aspect is the Aboriginal capacity to think and talk and sing and design in four dimensions today and in times past to the Dreaming. As a tool, this skill is an intellectual tool obtained from generational knowledge transferral and learning, but to Western persons it is the realm of geographical information systems (GIS) that is devised of the mid- to late-1900s inventory and modeling discourses. We know today that GIS is a digital software and hardware system designed to capture, store, manipulate, analyze, manage, and present spatial or geographic data, also referred to as geographic information science, the science underlying geographic concepts, applications, and systems (Steinitz 1993, 2014; Steinitz et al. 1976, 444–455). Regardless, as a tool GIS allows users to create interactive queries (user-created searches), analyze spatial information, edit data in maps, and present the results of all these operations.

That said, the realm of digital technologies has been slow to be embraced by Aboriginal communities and corporations across Australia. Many are navigating the Commonwealth and or state government expectations to articulate their voice, their vision of and for their Country, and to articulate the essence and aspirational management principles and actions for their Country. Such is constrained in Western requisites that it needs to be stated in text and diagram, expressed as static truths, and narrated as academic treatises capable of withstanding their tabling and defense in land-use litigation discourses within jurisdictions across Australia. A complexity is being able to squeeze a living tapestry culture, a dynamic system, of beliefs, lore, and protocols, that is constructed and protocoled by social responsibilities and principles into a western "codebook."

This complexity is evidenced in the 1970s with projects inside Uluṟu Kata Tjuṯa National Park (Australian Government, Department of Agriculture, Water and Environment 2016) and Kakadu National Park (Australian Government, Department of Agriculture, Water and Environment 2010) where dynamic land management principles were linked to defined pliable Indigenous-informed seasons resulted in a different management plan system. Plans of the 1990s for Kooyang (Framlingham Aboriginal Trust et al. 2004), Bunya (Bunya Mountains Aboriginal Trust et al. 2010), pushed the horizons on these attempts to construct culturally relevant management plans and statements. But they still struggled in obtaining Aboriginal community and corporation full acceptance as well as government endorsement. In the last ten years several innovations in expressing and narrating Country in plan have occurred, but most remain two dimensional textual essays washed with new graphics. Witness innovations for the Yawuru (Yawuru Registered Native Title Body Corporate 2011) and Murujuga (Murujuga Aboriginal Corporation 2016) communities, although the latter included digital applications, and the advances in Victoria with the Gunditjmara (Parks Victoria 2015) community and their recent Budj Bim Cultural Landscape (Australian Government, Department of Environment and Energy 2017) world heritage inscription. These precedent exemplars are holistic, and community "owned" in their drafting, expression, and graphics exposition. They all apply four-dimensional logic in three-dimensional documentation, and they are very extensive in their scope and intent. These contrast with conceptual Country Plans that are far shorter in pagination and vision ambit lacking the detail Western land use planners and managers are seeking in assisting "their" deliberations. The latter repeats the continued colonization power construct.

A second pattern, that has been occurring in the last twenty years, is the evolution of cultural mapping as a methodological approach (Poole 2003; Duxbury et al. 2015). This approach is different from the conventional Western communization of biogeography and cultural landscape attributes, like ecosystems, vegetation communities, slope angle, historic places, aspect, water system types, ridgelines, and so on. The former shifts quantitative socio-biogeography content to cultural-biogeography enabling qualitative values to be expressed and incorporated. Such mirrors the advent of cultural criteria in 1982 into the UNESCO World Heritage Convention when "values" become accepted as definitional attributes in defining place and human association to place. Cultural mapping has matured in Canada in First Nations applications, including the use of and acceptance of four-dimensional digital technologies. But it is in its formative stage in Australia. Hemisphere Design (2007) teased with graphical modeling in the 1990s, UDLA (2020), MudMap (Margetts et al. 2021), Nicholson et al. (2020, 1–19), and Brave & Curious (2021) have been exploring its use, as has also quietly several Aboriginal corporations seeking to construct a multilayered database geographic information system that are contemporary knowledge repositories of Country and not simply to serve its Western contrivance as a modeling/testing tool. The repository agenda is intentionally creating a "library" per Country but used the multilayered GIS system to map, express, document, and narrative information respectfully, discretely, and culturally appropriately within the cultural and spatial confines of their own Country. Such is increasingly being used as the basis of creating a Country Plan and forms the hidden appendix behind the Country Plan (Wadawurrung 2018; WTOAC 2020).

A third pattern, that has occurred in the last ten years, is the increasingly digital fluency of the younger generation. With an increasing passion, the youth of these Countries are becoming tech-savvy. Additionally, going into several corporation offices, behind the screened counters, will reveal a major interest in drone technologies by this same youth cohort that is now offering avenues of bringing aerial insights into the way Country is expressed, mapped, modeled, and monitored (Threadgold 2020).

WADAWURRUNG COUNTRY

With an understanding of Country and landscape as a multiplicity of layers, this section introduces Wadawurrung Country as a specific place. It considers the nature of Country Plan making, Cultural Mapping, and GIS through contemporary expectations across Australia.

Wadawurrung Country (Powell et al. 2019) resides within the Kulin Nation in the state of Victoria in Australia. Stretching from Beaufort and Ballarat in the north, encompassing over 10,000 km² (WTOAC 2020). It includes large parts of the Western Volcanic Plains and Werribee Plains as well as the Otway Ranges and the entire Geelong (Djillong) and Bellarine Peninsula Region. It also includes Country that, prior to the last glacial change 12,000 year ago, extends to the continental shelf as well as near half of the present Port Phillip Bay. Country does not discriminate between land and sea now and is [historically] Country before the "time of chaos" that flooded these tracts of landscape (Briggs 2008).

Figure 2.1. Wadawurrung Country.

Table 2.1. Wadawurrung Place Names. Abridged from Wadawurrung Aboriginal Corporation 2020, pp. 8.

English/Contemporary	Wadawurrung
Werribee	Weribbi
Torquay	Jan Juc
Aireys Inlet	Mangowak
Black Hill	Kareet Bareet
Geelong	Djilang
Beaufort	Yarram Yarram
Lake Burrumbeet	Burrumbeet
Skipton	Worram
You Yangs	Wurdi Youang
The Anakies	Anakie Youang

Within this Country is a mixture of geological landscapes and vegetation communities. To the far north are sedimentary uplift Great Dividing Ranges cloaked in dense wet sclerophyll *Eucalyptus* sp. forests that sweep down into the Western District's expansive grasslands growing on the largest and youngest volcanic plain on the globe. Between the Ballarat region and the Werribee region, in a north-south corridor, is an extensive sedimentary uplift raised corridor called the Brisbane Ranges that looks out over the latter and which is cloaked in a dense, dry sclerophyll *Eucalyptus* sp. forest. Both plains are derived from volcanic eruption lava from nearly 200,000–7,000 years ago (Wilkie et al. 2020, 403), with the mounts of Buninyong (Powell 2015a), Brown (Powell 2015b), You Yangs (Powell 2015c), and the Anakies (Tournier 2011, Ryan 2017) being their lava mouths and are to the Wadawurrung places of stories and ancestors. Anthropologically we know that the Wadawurrung have been here for some 40,000–60,000 years (Ecology and Heritage Partners 2019). Further, that their past generations witnessed the volcanic eruptions (Wilkie et al. 2020) on these plains that continued to erupt until 7,000 years ago at the same time as witnessing the "time of chaos" some 10,000–12,000 years ago when Port Philip Bay flooded, and their southern edges were submerged under Bass Strait (Briggs 2010, 8–11; Holdgate et al. 2011, 157–175). Their oral histories tell of these incidents.

The Otway's landscape mimics the Ranges but are wetter and denser, and the Geelong-Bellarine Peninsula hosts rolling sedimentary uplifted hills cloaked in a mix of open *Eucalyptus* sp.–dominated woodlands, *Melaleuca* sp., and gnarled *Eucalyptus* sp.–dominated heathlands and riverine threads, and undulating grasslands (Bellarine Catchment Network 2010a; Bellarine Catchment Network 2010b). The latter region, on an ancient dunal landscape, is fringed by coastal vegetation communities clinging to primary and secondary dunes encircling saline and freshwater threads, lakes, wetlands, and ponds, with associated cliff edges. The yellow flowering perennial grassland species, the Murnong (Gott 1983, 2–18), the icon of the Wadawurrung Traditional Owners Aboriginal Corporation (WTOAC) today, proliferated across much of these plains and open landscape, and was carefully cultivated by the Wadawurrung. Nestled within the landscape, in partnership with the Wadawurrung, was a vibrant community of predominately consuming herbaceous wildlife. The latter include eastern grey kangaroos, wallabies, together with a scatter of meat-eating predators like dingos, quolls, as well as a rich spectrum of permanent and seasonal avian species that swarmed along the riverine corridors, enjoying the endless rich grasslands, frequenting the now-listed RAMSAR saline and freshwater wetlands and lakes, and enjoined with the coastal dunes and shallows (Lunt et al. 1998; Lunt 1991, 56–66). Sustaining this vibrant vegetation and animal community was a long-established cultural burning regime that the Wadawurrung practiced across the region, using accepted cool burn practices and flexible seasonal time patterns in creating strategic mosaic patches and corridor burnings.

What is extant today is a mixed temperate, Mediterranean climatic landscape dominated by grasses and *Eucalyptus* sp., but some 12,000 years ago was a much drier landscape being influenced by active volcanoes like the Anakies (Anakie Youang).

On this landscape, the Wadawurrung were scattered in family or clan clusters, within sub-Country spaces, that each family or clan looked after for and on behalf of the larger Wadawurrung community. They stayed within the larger range of these sub-Countries, given available food and water resources (Nicholson et al. 2019, 2020, 1–19), but they met for annual cultural gatherings, rituals and ceremonies, trade and story and song exchange events, and to enable travel through traveling routes called Songlines (Powell et al. 1987, 44–84). Songlines littered the landscape like roads do today, but there was a greater cultural purpose to these than today. Within the landscape, clans and families could move according to their seasonal calendar, animal and food resource-sharing activities, and cultural incidents. Figure 2.2 depicts what we know as to their regular seasonal movement patterns. Understanding Indigenous seasons logic is a key to understanding precolonization Country

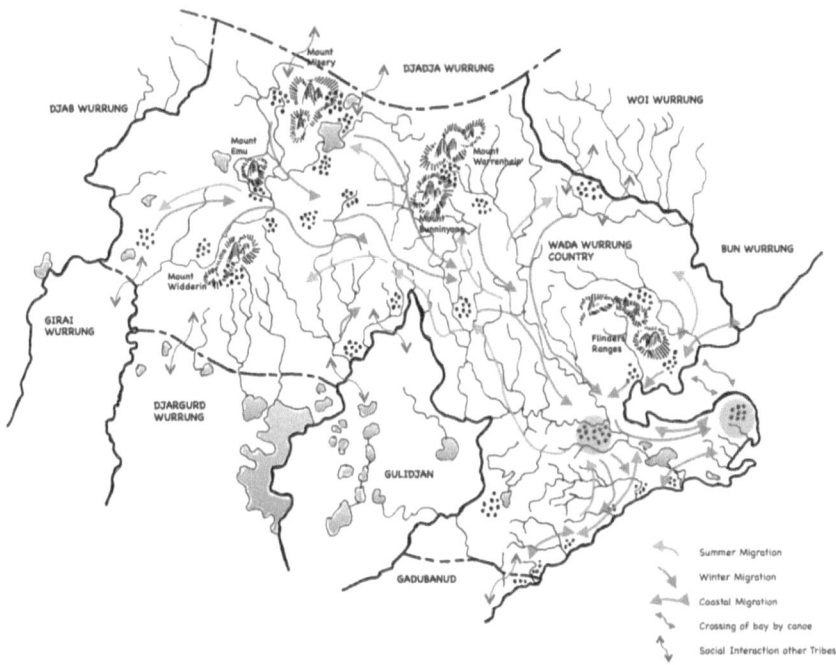

Figure 2.2. Wadawurrung Seasonal Movement Patterns.

relationships, apart from language, per Country community. Figure 2.3 depicts a reconstruction of a calendar pertinent to the Geelong-Bellarine region as distinct from the entire Wadawurrung Country (Powell et al. 1987). The core study area, discussed in this chapter, the southern tract of Wadawurrung Country, includes plains of Werribee, while sweeping around the western edge of Port Phillip forming today Geelong and the Bellarine Peninsula. Incised and twisting within this are the Barwon, Moorabool, and Werribee Rivers that circulate through the plains and hillslopes before originally exiting into the braided Birrarung Marr that ran north–south underneath the present Port Phillip Bay's waters before Bunjil shudder's (Briggs 2010, 8–11) altered their courses and character.

It has been said that some 2,000–5,000 Wadawurrung people lived across this landscape prior to the 1700s. Post 1700s, the onslaught of diseases, formal colonization invasions (Clark 1995) and dislocation, extensive vegetation removal, felling and firing and uncontrolled bushfires (Gammage 2011), and malnutrition and staple food dislocation (Gott 1982a, 59–67; 1982b, 13–17; 1983, 2–18; Gott 1985, 3–4; Nicholson and Jones 2020, 508–525), and

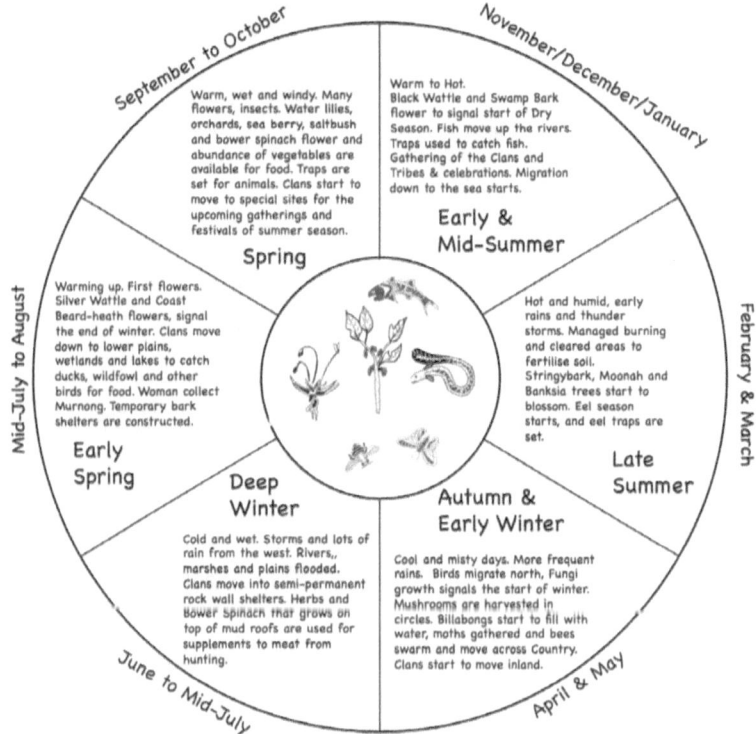

Figure 2.3: Seasons of the Geelong Region Wadawurrung Peoples.

reculturalization erased (Pascoe 2007; 2014) many Wadawurrung people, families, from their home Country, their regular and seasonal encampments, and their gathering locations. Bearing the hallmarks of dispossession, unconscious and conscious genocide, resulting in a major disrespectful invasion. This human chaos deeply affected and changed a landscape management regime that had largely been stable since the "time of chaos" after the settling of this climate change event, and this is without discussing the change wrought upon terrestrial, aquatic and marine animals, and avian species and their host vegetation communities (Briggs 2008; Jones et al. 2018, 402–417).

The descendants of the Wadawurrung are known and accepted descendants of apical ancestor, John Robinson (Robertson) (1846–1919), and his immediate descendants being Emily Hewitt, William Robertson, Mary Edith Hine, Victoria Alice Brannelly, Valentine Margaret Dalton, Mabel Violet Powell, Hector Norman Arthur Robinson, Ellen Rose King, and Thomas Joseph Russell Robinson, who are Wadawurrung according to Wadawurrung law and tradition. Their legal spokesperson, under the Aboriginal Heritage Act 2006 (Vic) is the Wadawurrung Traditional Owners Aboriginal Corporation (WTOAC) that is seeking to care for Country, through contemporary mechanisms, policies, and activities. The Corporation is the Registered Aboriginal Party (RAP) for Wadawurrung Country with the statutory authority for the management of Aboriginal heritage values and culture, under the Act.

POSTCONTACT KNOWLEDGE AND NARRATIVES OF WADAWURRUNG COUNTRY: 1830S–1870S

A significant change to the contact experienced by the Wadawurrung and other Kulin Nations with non-Aboriginal people occurred during the 1830s. This change instigated the genocide of Kulin Nation peoples, and the colonization and dispossession of their Country.

With the pastoral expansion in Van Diemen's Land (now Tasmania) from 1821, the colonial population increase dramatically reduced available productive land in addition to the struggle for possession of Country, called The Black War (Ryan 2008). Interested entrepreneurial investigation to seek agricultural possibilities and expanding their interest outside of Van Diemen's Land, John Batman along with Joseph Tice Gellibrand attempted in 1827 to obtain a land grant at Port Phillip (Roberts 2007, 5). Their failed attempt led to the establishment of the Geelong and the Dutigalla Association, later known as Port Phillip Association, led by Batman, Gellibrand, Wedge, and Charles Swanston (Library Council of Victoria 1982). Where land grants had failed the Port Phillip Association aimed to settle Port Phillip through the "purchased" cession of land by the Kulin Nation peoples (Library Council of Victoria 1982; Roberts 2007, 6).

For the Wadawurrung, uninvited invasion occurred at Indented Head, on the Bellarine Peninsula, in 1834. It also occurred by errant human and sheep trampling, and colonized "acquisition" of their Country unbeknownst to them on the edge of Merri Merri Creek in Wurundjeri Country in 1834. Wadawurrung and Wurundjeri Country, including the Western Volcanic Plains, Werribee Plains and Bellarine Peninsula, was surveyed and divided into seventeen shares by the Association. The Association's corporate aim was a significant change to the contact previously experienced by Kulin Nation peoples. It was this change that led to their genocide and dispossession of *Country*, resulting in the loss of history, culture and agriculture and vegetation knowledge.

During this postcontact period, the newfound interest in the region generated through Port Phillip Association's reports of rich agricultural land led to the invasion by non-Aboriginal people on Country. These colonists came from diverse social backgrounds, and later characterized as "Van Diemen's Landers" who had followed the likes of Batman; and the "Overlanders" from the likes of Sydney; and the "Immigrants Settlers" from Britain (Christie 1979, 29–30). Today these colonial family nomenclatures are littered across Victoria in place names, while the Aboriginal language and clan names are almost hidden.

Notable "Van Diemen's Landers" were pastoralist George Russell and the Manifold brothers Thomas, John and Peter. Russell had heard of the fine country discovered at Port Phillip by Batman and John Fawkner, and in late March 1836 left the *kanamaluka* (River Tamar) to explore Port Phillip (Brown 1935, 75–78). Using a tracing of what was then known as "Wedge's Map," Russell ventured onto Wadawurrung Country, exploring the land and waterways around the *wirribi-yaluk* (Werribee River), *mooroobull* (Moorabool River), Barwon River, Leigh River, and Native Creek before returning to Van Diemen's Land (Brown 1935, 82–85). On his return, Russell went on to meet the Manifold family, when in 1836 Thomas Manifold had kindly offered him a seat in his boat to Launceston (Brown 1935, 103). It was in October of the same year Russell returned to Port Phillip, securing suitable land to settle the headquarters of the Clyde Company. The Clyde Company established on *mooroobull* at Batesford in 1839 and were neighbours with the Manifolds. The first sale of land in Port Phillip District saw Manifold brothers Thomas, John, and Peter, relocate, and settle "Purrumbete," "Weerite" in 1839. While the Clyde Company moved their operation to "Golf Hill Estate" on the Leigh River at Shelford, later extending to *terinallum*, near Mount Elephant in 1846 and *mawallok*, near Stockyard Hill in 1847 (Brown 1935, 180–82; 248–250). Many of the "Van Diemen Landers" who became prominent on *Wadawurrung Country* were prominent in the government and settlers Black War on Tasmania Aborigines in 1821–1831 (Ryan 2008). The subsequent invasion, genocide, and dispossession on Wadawurrung Country is not unprecedented considering the death and destruction that they had left in Van Diemen's Land.

A noteworthy "Overlander" is pastoralist Thomas Chirnside. In 1840 Andrew Chirnside assisted brother Thomas to bring a flock of 700 merinos overland from Murrumbidgee to Loddon (Hone 1969). The brothers extended their land holding throughout the Port Phillip District; furthering them into the Western District including stations at "Mokanger," "Woady Yallock," "Curnong," and "Carranballac" (Hone 1969, Jones 1993). It was in the 1850s they began increasing their land holdings around *wirribi-yaluk* (Werribee), purchasing the property named Werribee Lower, previously owned by Joseph Gellibrand (Wyndham City Council 2020). In 1853, they would acquire "Werribee Park" from the Wedge family. The property had been established by Edward Wedge in 1836, John's brother (Heritage Victoria 2012). After the 1852 *wirribi-yaluk*, Werribee River floods that had claimed the life of Edward, his wife Lucy, and their daughter, those remaining in the Wedge family transferred the land to their neighbor Thomas Chirnside (Heritage Victoria 2012). It was determination for profitable and productive land like those perused by the Chirnsides that resulted in the loss of Aboriginal culture, agriculture, and vegetation knowledge through the dispossession of Country.

An unusual "Immigrant Settler" for their time but a significant name was Drysdale. Holding only a leasehold in Scotland, new land at Port Phillip offered those with little capital something they could not afford in Scotland— land (Roberts 2009, 7). In 1839 with modest capital, farm manager experience, and social connections already in Port Phillip, at forty-seven years of age, Miss Anne Drysdale sailed for Port Phillip from Scotland (Roberts 2009, 8). She invited Miss Caroline Newcomb to partner with her on a squatting run, "Borronggoop," beside the Barwon River in Geelong (Roberts 2009, 9). Like most pastoralists she was determined for profitable land, and from a modest beginning to prosperity with Newcomb they would move to "Coriyule" on the Bellarine Peninsula (Roberts 2009, 10). It was at "Coriyule" that they reduced their pastoral connection and took to horse breeding and growing crops (Roberts 2009, 10). The determination for prosperity by the colonist was not advantageous for the Kulin Nation peoples. Ultimately this saw their genocide and dispossession of land that resulted in the loss of Aboriginal culture, agriculture, and vegetation knowledge, which we are still striving to get back today.

JOHN HELDER WEDGE AND HIS TRAVELS AND NARRATIVES

A key agent and facilitation in this "exploration" of Wadawurrung Country was John Helder Wedge. Wedge was born in February 1793 in England (Crawford et al. 1962, xi). Wedge was the younger son of Charles Wedge from Shudy Camps, Cambridge, England (Stancombe 1967). Charles was a surveyor who assisted civil engineer John Rennie in the construction of the English canal system and was also engaged in the great survey of England conducted in the 1820s (Stancombe 1967). Wedge learned the fundamentals of surveying from his father, Charles, assisting him in work which would later be his profession in Van Diemen's Land (Crawford 1962). Though Charles was also known for practicing modern farm practices, during his youth Wedge assisted his elder brother, Edward David Wedge, in farm operations (Crawford et al. 1962, xi).

The postwar depression in agriculture in England motivated Wedge and Edward to emigrate to Van Diemen's Land (Crawford et al. 1962, xi). Wedge's five-month journey commenced in 1823 with Edward, £1000 of sawmill machinery and twenty-three merino sheep, including two rams (Crawford et al. 1962, xi). On arriving Wedge commenced an appointment at the colonial Survey Department as assistant surveyor (Crawford et al. 1962, xi). With the wealth Wedge had brought to Van Diemen's Land, he was granted 607 ha near Perth, which he called "Leighlands" (Crawford et al. 1962, xi).

During 1824, Wedge's Tasmanian journals indicate that he was engaged in field surveying in the town of Brighton (Crawford et al. 1962, xii). Wedge's brother Edward was granted 607 ha in the parish of Beverley, next to John Batman at Ben Lomond in 1827 (Crawford et al. 1962, xv). It was during this time that Batman and Wedge became good friends. This saw them accompany Governor Arthur on an expedition to the George's River (Crawford et al. 1962, xv). A letter from Wedge to his father Charles in England noted that in 1833 he, John Batman, and John Glover—an artist—would journey to the summit of Ben Lomond (Crawford et al. 1962, xv). It was later this same year that Wedge was engaged to survey along the Ben Lomond Rivulet. Wedge led a group, in a large expedition organized by Surveyor-General George Frankland to explore the land between the Derwent, Gordon, and Huon Rivers during the early part of 1835 before his resignation from the Survey Department (Crawford et al. 1962, xvi; Stancombe 1967). With stories of oceans of grass rolling for hundreds of miles by Hovell and Hume (Pritchard 2020), Wedge along with Batman had been motivated to explore across Bass Strait for suitable land for increasing flock and Wedge's dissatisfaction with the nepotism of the Colonial Office (Crawford et al. 1962, xv; Stancombe 1967).

Wedge was then appointed as "surveyor" to the Port Phillip Association, to explore, survey, and map the Port Phillip treaty area, dividing it into seventeen shares (Library Council of Victoria 1982). The Association were a group of interested businessmen established to investigate and colonize land at Port Phillip, alleged by Batman to have been negotiated under a treaty with the Kulin Nation peoples (Brown 1989). The three manuscripts that detail the expedition to Port Phillip that resulted in the invasion and colonization of Wadawurrung Country and the broader Kulin Nation are held at the State Library of Victoria (Brown 1989). These manuscripts are penned by Batman, Wedge, and Andrew (alias William) Todd, Batman's recorder.

Peter Alsop, in analyzing Wedge's field book, traced and mapped Wedge's supposed routes for August and September of 1835. These supposed routes were divided into three journeys.

The first journey in early to mid-August 1835 includes the localities now known as St. Leonards, Indented Heads, Drysdale, Portarlington, Connewarre, and Ocean Grove before returning to St. Leonards (Brown 1989). It was around this time that Governor Bourke issued a proclamation declaring "Batman's Treaty" void and the Port Phillip Association trespassers on Crown Land (Roberts 2007, 8).

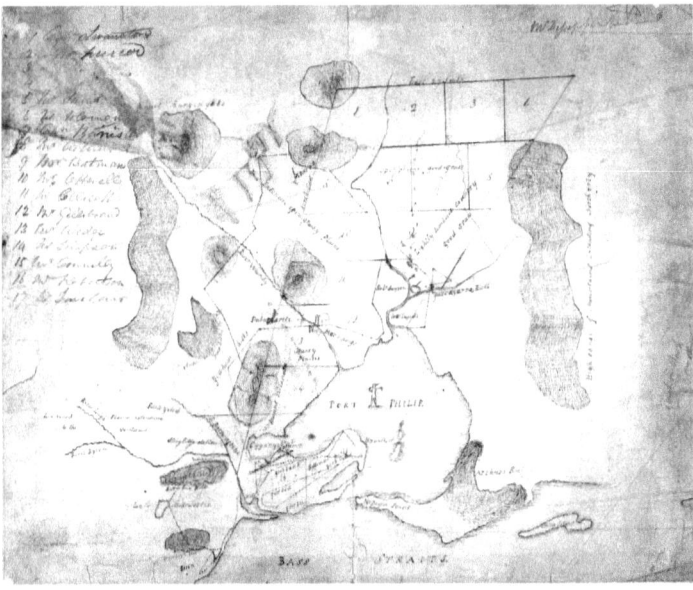

Figure 2.4: John Helder Wedge—Plan of The Port Phillip District, 1835.

The second journey, in mid-August 1835, includes the localities now known as Geelong, Fyansford, Barrabool Hills, Waurn Ponds, Moriac, Lake Modewarre, Point Roadnight, Point Addis, Bells Beach, Torquay, Mount Duneed, and Connewarre (Brown 1989). Wedge used his knowledge drawn from farming to depict the landscape he surveyed. During this journey Wedge scribed in his field book:

> . . . then passed over some low rises to a water course called by the natives Byourac [sic.] this was more thickly timbered gum trees prevailing—good grass continued up to this place, the soil being more tenacious after crossing the watercourse the rises increase in altitude the soiled and grass not so good. . . . (Wedge 1835, 68)

The final journey, attributed by Alsop, was in late August to early September to the localities now known as Geelong, Lara, Balliang, Melton, Diggers Rest, Keilor, Essendon, and Melbourne returning through Newport, Laverton, Werribee, Little River, the You Yangs, Lara, and Lovely Banks. Wedge describes the landscape near Anakie Youang in his field book as:

> . . . intending to ascend one of the emmenences which for the cluster of hills [Anikai Youwham] but learning from the two native youths that were with me that there was no water in that direction I altered my course and passed over a ridge of hills run from Mt Gowwham open plains and continue nearly to the ridge they are quite destitute of timber and covered with grass but not of that luxuriance that I should expected for the soil [sic.]. (Wedge 1835, 61)

Figure 2.5: John Helder Wedge—Barrabull Hills August 20, 1835. Sketch in John Helder Wedge Field Book 1835–1836, Image 88.

Figure 2.6: Charles Norton—Near Fyansford 1846.

Wedge's mapping, coding, triangulating, cross-referencing, and scrawling names on his hardcopy plans and, in his diaries, proclaiming this patch of *terra nullius* to be Port Phillip District and renaming all within his traverses with a mixture of largely anglicized nomenclature or abridged Wadawurrung nomenclature albeit reinterpreted and spoken by non-Wadawurrung Indigenous surveying guides, less some assistance by escaped convict Buckley (Courtney 2020). In successive years huge flocks of sheep were released guided by Wedge's survey maps, and several European squatters and émigré's narrated the resources and qualities of this landscape in their diaries scrawls, or sketched and painted abstract images and idealized scenes.

A struggle commenced in 1836 between the Association and the Colonial Office, with the former seeking recognition of their land claims (Crawford et al. 1962, xvii). In late 1836 Governor Bourke took Batman's Port Phillip under government control (Brown 1989). The Association was disbanded, with most members selling out of their interests (Crawford et al. 1962, xvii). Wedge, however, held his interest in Port Phillip until 1854, after extending this in 1837. Wedge added to his interest during the first auctions of land purchasing two land blocks extending his run at Werribee (Crawford et al. 1962, xix).

From 1838–1843 Wedge visited England, and on the death of his father, Charles, he returned to Tasmania (Stancombe 1967). In 1843, he married

Maria Medland Wills, but within a year she died in childbirth (Stancombe 1967). Wedge was appointed in 1846 to manage the farms financed by the Christ's College at Bishopsbourne in Hobart (Stancombe 1967). In 1855, Wedge was elected to the Tasmanian upper house Legislative Council for the District of Morven until his departure from parliament in 1868 (Stancombe 1967). Wedge had built a home, "Medlands," on the River Forth located in northwest Tasmania in 1865. It was here he died on November 22, 1872 (Stancombe 1967).

With this historical context, a range of primary sources document the key colonial visitors to Wadawurrung Country, in the Werribee-Geelong-Bellarine region in the 1830s to 1870s. The primary recording evidence is diaries and paintings/sketches that offer detailed descriptions of this landscape, which can be cross-tabulated against Wedge's travels and diary entries.

Wedge's diaries and maps coupled with this analysis form the basis and key understanding of the original cultural living landscape and its biogeographical construction. To date Wedge's surveys, maps, and diaries have been little calibrated (both quantitatively and qualitatively) to test the veracity of the pre-1800 vegetation hypothesis (Carr et al. 2008; 2006) used by the state Department of Environment Land Water and Planning (DELWP). In addition, mapping and understanding the actual landscape mosaic in 1835–1836 when he first sighted and traversed this tract has not been captured and analyzed in GIS mapping.

DIGITAL TOOLS REPRESENTING PLACE FOR WADAWURRUNG COUNTRY

Clashes between Indigenous and colonial concepts of "place" demonstrate the need for an innovative common space where storytelling can take place, storytelling that is inclusive and instructive and reflective of a journey partnership with the Indigenous participant. The use of GIS tools provides a digital space that can facilitate storytelling that is recognizable to many who need to understand place when working with Country. By tracing the path of dispossession through time, correlating spatial references with colonist's writings and artwork in GIS mapping it would enable a description of the landscape at the point of Wadawurrung loss while enabling a connection to "place."

GIS can then operate as a tool for the Wadawurrung to facilitate culturally appropriate management of Country in partnership with government authorities and those responsible for land use [or rather, custodianship] planning management. The continuity of the land use in the region is incorrectly narrated, however placing the colonizers in the larger story of Country and

Figure 2.7. GIS-based layers of the Vegetation and Water Systems in the Research Area.

landscape redefines the narrative and begins to open a dialogue of understanding of place over longer timescales. Taking Wedge's detailed observations and placing them in context of Country through the use of GIS builds local context, beginning to build knowledge that helps us find "place" and understand its cultural nuances.

REFERENCES

Australia Government. 2010. *Uluru-Kata Tjuta National Park Management Plan 2010–2020.* Department of Agriculture, Water and Environment (Canberra ACT).

———. 2017. *Budj Bim Cultural Landscape World Heritage Nomination.* Australian Department of the Environment and Energy (Canberra, ACT).
Australia Government Department of Agriculture, Water, and Environment. 2016. *Kakadu National Park Management Plan 2016–2026.* (Canberra, ACT).
Australia ICOMOS. 2013. *The Burra Charter: The Australia ICOMOS Charter for Places of Cultural Significance.* International Council on Monuments and Sites (Canberra, ACT).
Bellarine Catchment Network. 2010. *Coastal Plants of the Bellarine Peninsula.* Bellarine Catchment Network (Drysdale, Victoria).
———. 2010. *Inland Plants of the Bellarine Peninsula. Bellarine Catchment Network* (Drysdale, Vic).
Brave & Curious. 2021. "We Are Brave and Curious." https://www.braveandcurious.com.au/.
Briggs, C. 2008. *The Journey Cycles of the Boonwurrung: Stories with Boonwurrung Language.* Melbourne Victoria: Victorian Aboriginal Corporation for Languages.
———. 2010. "Boonwurrung: The Filling of the Bay—The Time of Chaos." In *Indigenous Creation Stories of the Kulin Nation,* 8–11. Melbourne: Arts Victoria.
Brown, P. L. 1935. *The Narrative of George Russell of Golf Hill with Russellania and Selected Papers.* London: Oxford University Press.
———. 1989. *The Todd Journal 1835: Andrew Alias William Todd John Batman's Recorder and His Indented Heads Journal 1835.* Geelong: The Geelong Historical Society.
Bunya Mountains Elders Council and Burnett Mary Regional Group. 2010. *Bonye Bu'ru Booburrgan Ngmmunge: Bunya Mountains Aboriginal Aspirations and Caring for Country Plan.* Markwell Consulting & Bunya Mountains Elders Council and Burnett Mary Regional Group (Canungra QLD).
Carr, G. W., L. A. Ashby, J. S. Kershaw, D. Frood, and N. G. Rosengren. 2008. *Mount Eccles Lava Flow Botanical Management Plan: Field Survey and Analysis* (unpublished). Ecology Australia (Fairfield).
Carr, G. W., D. Frood, N. R. Roberts, and N. Rosengren. 2006. *Mt Eccles Lava Flow Botanical Management Plan: Literature Review.* Ecology Australia (Prepared for Winda-Mara Aboriginal Corporation. Fairfield).
Christie, M. 1979. *Aboriginals in Colonial Victoria 1835–1836.* Sydney: Sydney University Press.
Clark, I. D. 1995. *Scars in the Landscape: a Register of Massacre Sites in Western Victoria, 1803–1859.* Canberra, ACT: Aboriginal Studies Press for the Australian Institute of Aboriginal and Torres Strait Islander Studies.
CoG. 2019. "City of Geelong." Public Images. http://www.geelongaustralia.com.au/common/public/images/inpage/kaap02.jpg.
Courtney, A. 2020. *The Ghost and the Bounty Hunter: William Buckley, John Batman and the Theft of Kulin Country.* Sydney, NSW: HarperCollins Publishers.
Crawford, Justice, Ellis W.F., and Stancombe. G.H. 1962. *The Diaries of John Helder Wedge 1824–1835.* Hobart: The Royal Society of Tasmania.

Duxbury, N., Garrett-Petts W.F., and D. MacLennan. 2015. "Cultural Mapping as Cultural Inquiry." In *Advances in Research Methods*, edited by N. Duxbury, Garrett-Petts W.F. and D. MacLennan. Routledge.

Ecology and Heritage Partners. 2019. *Drysdale Bypass, Jetty Road to Whitcombes, Drysdale, Victoria: Aboriginal Cultural Heritage Management Plan CHMP 13839*. Ecology and Heritage Partners (Ascot Vale, Victoria).

Framlingham Aboriginal Trust and Winda Mara Aboriginal Corporation. 2004. *Kooyang Sea Country Plan*. FAT & WMAC (Warrnambool, Victoria). http://www.environment.gov.au/resource/kooyang-sea-country-plan.

Gammage, B. 2011. *The Biggest Estate on Earth: How Aborigines Made Australia*. Crows Nest NSW: Allen & Unwin

Gott, B. 1982. "Ecology of Root Use by the Aborigines of Southern Australia." *Archaeology in Oceania* 17: 59–67.

———. 1982. "*Kunzea pomifera*–Dawson's Nut." *The Artefact* 7 (1–2): 13–17.

———. 1983. "Murnong—*Microseris scapigera*: a Study of a Staple Food of Victorian Aborigines." *Australian Aboriginal Studies* 2: 2–18.

———. 1985. "'Plants' Mentioned in Dawson's Australian Aborigines." *The Artefact* 10: 3–14.

Gott, J., and B. Conran. 1991. *Victorian Koorie Plants: Some Plants Used by Victorian Koories for Food, Fibre, Medicines and Implements*. Yangennanock Women's Group.

Hemisphere Design. 2007. *Kaurna Cultural Heritage Survey July 2007*. Hemisphere Design (Adelaide, SA).

Heritage Victoria (HV). 2012. *Victorian Heritage Database Place—Werribee Park*. VHR No H1613: Heritage Victoria.

Holdgate, G. R., B. Wagstaff, and S. J. Gallagher. 2011. "Did Port Phillip Bay Nearly Dry Up between 2800 and 1000 Cal. Yr BP? Bay Floor Channelling Evidence, Seismic and Core Dating." *Australian Journal of Earth Sciences* 58 (2): 157–75.

Hone, J. A. 2020. "Chirnside, Thomas 1815–1887." *Australian Dictionary of Biography, Nation Centre of Biography*. Australian National University. Accessed September 28, 2020. http://adb.anu.edu.au/biography/chirnside-thomas-3203.

Jones, D. S. 1993. "Cockatoo (*Chinna junnak cha knœk grugidj*): A Quest for Landscape Meaning in the Western District, Victoria, Australia (unpublished)." PhD, University of Pennsylvania.

Jones, D. S., K. Alder, S. Bhatnagar, C. Cooke, J. Dearnaley, M. Diaz, H. Iida, Nair A. Madhavan, S-L. McMahon, M. Nicholson, G. Pocock, B. Powell, G. Powell, S. G. Rahurkar, S. Ryan, N. Sharma, Y. Su, S. V. Wagh, and Yapa O. Appuhamillage. 2021. "Country." In *Learning Country in Landscape Architecture Indigenous Knowledge Systems, Respect and Appreciation*, edited by Jones D. S. London, UK: Palgrave Macmillan.

Jones, D. S., P. B. Roös, J. Dearnaley, H. Threadgold, M. Nicholson, R. Wissing, D. Berghofer, R. Buggy, D. Low Choy, P. A. Clarke, S. Serrao-Neumann, G. Kitson, S. Ryan, B. Powell, G. Powell, and M. G. Kennedy. 2018. "ReCrafting Urban Climate Change Resilience Understandings—Learning from Australian Indigenous

Cultures in Biophilia Smart Resilience." 55th International Federation of Landscape Architects World Congress, Marina Bay, Singapore, 18–21 May 2018.

Library Council of Victoria. 1982. *Trespassers and Intruders: The Port Phillip Association and the Founding of Melbourne*. Melbourne, Victoria: Library Council of Victoria.

Lunt, I. 1991. "Management of Remnant Lowland Grasslands and Grassy Woodlands for Nature Conservation: A Review." *Victorian Naturalist* 108 (3): 56–66.

Lunt, I., T. Barlow, and J. Ross. 1998. *Plains Wandering: Exploring the Grassy Plains of South-Eastern Australia.* Melbourne, Victoria: Victorian National Parks Association & the Trust for Nature.

Margetts, V., and B. Pigram. 2021. "Culture in Design." In *Everyone's Knowledge in Country: Yurlendj-nganjin*, edited by D. S. Jones and D. Low Choy. Newcastle upon Tyne, UK: Cambridge Scholars Publishing.

McHarg, I. L. *Design with Nature*. 1969. Garden City, N.Y. American Museum of Natural History: Natural History Press.

Meinig, D. W. 1979. "The Beholding Eye: Ten Versions of the Same Scene." In *The Interpretation of Ordinary Landscapes: Geographical Essays*. edited by D. W. Meinig and J. B. Jackson. New York: Oxford University Press.

Mitchell, T. L. 1965. *Three Expeditions into the Interior of Eastern Australia, with Descriptions of the Recently Explored Region of Australia Felix, and the Present Colony of New South Wales* (2 Vols). London, UK: T & W Boone 1839; facsimile edition, Adelaide, SA Libraries Board of South Australia

Mulvaney, J., and J. Kamminga. 1999. *The Prehistory of Australia*. Sydney: Allen and Unwin.

Murujuga Aboriginal Corporation. 2016. *Ngaayintharri Gumawarni Ngurrangga: We All Come Together on This Country—Murujuga Cultural Management Plan 2016*. Murujuga Aboriginal Corporation (Karratha, WA).

Nicholson, M., and D. S. Jones. 2021. "Wurundjeri-al Biik-u (Wurundjeri Country), Mag-golee (Place), Murrup (Spirit) and Ker-up-non (People)." In *Handbook on Historic Urban Landscapes of the Asia-Pacific*, edited by K. D. Silva, 508–25. Abingdon, UK: Routledge.

Nicholson, M., G. Romanis, I. Paton, and D. S. Jones. 2019. *North Gardens Sculpture Park Landscape Master Plan*. School of Architecture & Built Environment, Deakin University (Geelong Victoria).

Nicholson, M., G. Romanis, I. Paton, D. S. Jones, K. Gerritsen, and G. Powell. 2020. "'Unnamed as Yet': Putting Wadawurrung Meaning into the North Gardens Landscape of Ballarat." *UNESCO Observatory E-Journal Multi-disciplinary Research in the Arts* 6 (1): 1–19.

Parks Victoria (PV), and Gunditj Mirring Traditional Owners Aboriginal Corporation (GMTOAC) and Winda Mara Aboriginal Corporation (WMAC) 2015. *Ngootyoong Gunditj Ngootyoong Mara South West Management Plan*. Melbourne, Victoria: Parks Victoria.

Pascoe, B. 2007. *Convincing Ground: Learning to Fall in Love with Your Country*. Canberra, ACT: Aboriginal Studies Press.

———. 2014. *Dark Emu: Black Seeds Agriculture or Accident?* Broome Western Australia: Magabala Books.
Poole, P. Cultural Mapping. 2003. "Cultural Mapping and Indigenous People." UNESCO. http://www.ibcperu.org/doc/isis/11953.pdf.
Powell, B. 2015. "Wadawurrung language: Bonan Youang." Open ABC. Accessed January 1, 2018. https://open.abc.net.au/explore/86782.
———. 2015. "Wadawurrung Language: Kareet Bareet." Open ABC. Accessed January 1, 2018. https://open.abc.net.au/explore/86715.
———. 2015. "Wadawurrung Language: Wurdi Youang." Open ABC. Accessed January 1, 2018. https://open.abc.net.au/explore/86719.
Powell, G., and D. S. Jones. 2018. "Kim-barne Wadawurrung Tabayl: You Are in Wadawurrung Country, Kerb." *Journal of Landscape Architecture* 26: 22–25.
Powell, (Uncle) B., D. (Tandop) Tournier, D.S. Jones, and P.B. Roös. 2019. "Welcome to Wadawurrung Country." In *Geelong's Changing Landscape: Ecology, Development and Conservation*, edited by D. S. Jones and P. B. Roös, 44–84. Melbourne, Victoria: CSIRO Publishing.
Pritchard, L. 2020. *Hume & Hovell Journey of Discovery to Port Phillip 1824–1825* (unpublished).
Pullin, R. 2010. "Eugene von Guerard: A Journey through Victoria's Western Plains." In *Designing Place: An Archaeology of the Western District*, edited by L. Byrne, H. Edquist and L. Vaughan, 112–25. Melbourne Victoria: Melbourne Books.
———. 2011. *Eugene von Guérard: Nature Revealed*. Melbourne, Victoria: National Gallery of Victoria.
Roberts, B. 1993. *The Cultural Heritage of the Barwon River*. Barwon Water (unpublished).
———. 2009. *Miss D and Miss N, an Extraordinary Partnership: The Diary of Anne Drysdale*. Melbourne: Australian Scholary Publishing.
Roberts, H. 2007. *Paved with Good Intentions: Terra Nullis, Aboriginal Land Rights and Settler-Colonial Law*. Act: Halstead Press.
Roös, P. B. 2017. "Regenerative-Adaptive Design for Coastal Settlements: A Pattern Language Approach to Future Resilience." PhD, School of Architecture & Built Environment, Deakin University.
Rose, D. B. 1996. *Nourishing Terrains: Australian Aboriginal Views of Landscape and Wilderness*. Canberra: Australian Heritage Commission.
Ryan, L. 2008. "List of Multiple Killings of Aborigines in Tasmania: 1804–1835." *SciencePo*. Accessed April 24, 2020. http://bo-k2s.sciences-po.fr/mass-violence-war-massacre-resistance/fr/document/list-multiple-killings-aborigines-tasmania-1804-1835.
Ryan, S. 2016. "Creative Places in South West Victoria." SRL733, School of Architecture & Built Environment, Deakin University.
———. 2017. "Conserving Our Indigenous Landscapes to Acknowledge Our Country's History." SRR711, School of Architecture & Built Environment, Deakin University.
SLV-1. 2019. "State Library Victoria." Public Images. Government of Victoria. https://viewer.slv.vic.gov.au/?entity=IE20234697&mode=browse

SLV-2. 2019. "State Library Victoria." Public Images. Government of Victoria. https://viewer.slv.vic.gov.au/?entity=IE20233586&file=FL20237490&mode=browse.

SLV-3. 2019. "State Library Victoria." Public Images. Government of Victoria. https://viewer.slv.vic.gov.au/?entity=IE457242&mode=browse.

Stancombe, G. H. 1967. "Wedge, John Helder (1793–1872)." *Australian Dictionary of Biography, National Centre of Biography.* Australian National University. Accessed October 5, 2020. http://adb.anu.edu.au/biography/wedge-john-helder-2778/text3951.

Steinitz, C. 1993. *Geographical Information Systems: A Personal Historical Perspective, the Framework for a Recent Project, and Some Questions for the Future.* GIS Europe.

Steinitz, C., P. Parker, and L. Jordan. 1976. "Hand-Drawn Overlays: Their History and Prospective Use." *Landscape Architecture* 9: 444–55.

Threadgold, H. 2020. "'What the Stones Tell Us': Gulijan Living Spaces and Landscapes." PhD, Deakin University.

Tournier, D. 2011. "Wathaurong Creation Stories." DVD, *The Three Sisters, Bundjil and the Making of the Land, Peerreedek the Platypus, Parwang the Magpie, Kangaroo and Wombat.* Geelong North: Wathaurong Aboriginal Co-operative.

UDLA. 2020. *UWA Cultural Heritage Mapping: Crawley Precinct.* UDLA (Fremantle, WA).

Wadawurrung. 2018. *Wadawurrung Country of the Victorian Volcanic Plains.* Ballarat, Victoria: Wadawurrung (Wathaurung Aboriginal Corporation).

Wedge, J. H. 1835. *Field Book 1835–1836* (9.5 x 19.4 cm). Port Phillip Papers Digitising Project; Port Phillip Association & Victorian Vision. State Library of Victoria, Melbourne, Victoria.

Wilkie, B., F. Cahir, and I. D. Clark. 2020. "Volcanism in Aboriginal Australian Oral Traditions: Ethnographic Evidence from the Newer Volcanics Province." *Journal of Volcanology and Geothermal Research* 403: 106999.

WTOAC. 2020. *Paleert Tjaara Dja—Let's Make Country Good Together 2020–2030: Wadawurrung Country Plan.* (Ballarat, Victoria: Wadawurrung Traditional Owners Aboriginal Corporation).

———. 2020. "Paleert Tjaara Dja, Healthy Country Plan 2020–2030." Wadawurrung Traditional Owners Aboriginal Corporation. https://www.wadawurrung.org.au/healthy-country-plan-video.

Wyndham City Council. 2020. "Wyndham Our Story—The Chirnside and Werribee Park." Wyndham City Council. Accessed September 26, 2018. http://wyndhamhistory.net.au/exhibits/show/topic-3---who-were-the-settler/thechirnsidesandwerribeepark.

Yawuru, Registered Native Title Body Corporate. 2011. *Walyjala-jala buru jayida jarringgun buru Nyamba Yawuru ngan-ga mirli mirli· Planning for the Future*: Yawuru Cultural Management Plan. Yawuru RNTBC (Broome WA).

Chapter Three

Tech for TEK

The Value of GIS Systems in Sustainable Community Planning and Indigenous Land Protection Initiatives

Shahreen Shehwar

In 2006, the Canada Mortgage and Housing Corporation conducted a case study on the Cree Nation of Mistissini, who were the first Indigenous community to use GIS mapping for community planning. The Mistissini were using GIS mapping for all types of community planning initiatives, including housing developments, infrastructure planning, and natural resource management. From the case study, it was evident that GIS mapping had been a capacity-building tool for the Mistissini, who were able to learn GIS skills and gain formal GIS training. This allowed them to undertake several community-building initiatives. For example, the community was able to use this tool to ensure that sewage outlets and drainage channels were not constructed too close to potable water intakes, protecting the water quality in the community (Sutton 2006).

Around this time, several studies and articles began to emerge that documented the benefits of participatory GIS. Participatory GIS (PGIS) is defined as a participatory approach to spatial planning, spatial information, and communications management (Rambaldi, et al. 2005). It is geared toward "community empowerment through measured, demand-driven, user-friendly, and integrated applications of geo-spatial technologies" (ibid). While early GIS was exclusive and unaffordable, used mainly by senior governments and a few skilled experts (Craig, Harris and Weiner 2002), PGIS was different because it was accessible and had the potential to profoundly impact community empowerment and produce social change for marginalized communities.

This is because PGIS provided communities the tools to generate and control the use of "sensitive spatial information" and "protect traditional knowledge and wisdom from external exploitation" (Rambaldi et al. 2005).

PGIS was aspirational and its potential did not go unrecognized by First Nations communities in Canada. Today, many First Nations communities are using GIS technologies in ways that extend beyond natural resource management purposes. For instance, since January 2011, the Matawa First Nations, a tribal council representing nine First Nations in northwestern Ontario, have been using GIS technology to store traditional knowledge data, "which is then applied to discussions with mining companies regarding potential claims in its territory" (Kelly 2012). Here, the Matawa First Nations are using GIS technologies to mobilize their traditional knowledge, so they can preserve their communities and engage with mining companies, and others, who have an interest in the territory. However, there are still many First Nations communities without access to these tools, and this is resulting in communication barriers with policymakers and gaps in data collection and preservation. In a study on the perspectives of Canadian Indigenous peoples toward environmental assessments, the highly technical nature of environmental assessments, which involved the use of GIS, was presented as a communication barrier for many First Nations communities, who felt excluded in the environmental assessments process (Booth and Skelton 2011). Additionally, in the absence of data, both the communities and the government were unable "to predict the consequences of industrial developments" (ibid). This was an impediment for natural resource management, as it begs the question, how could policymakers make decisions without proper data? Considering that First Nations communities have so much crucial knowledge of their natural surroundings, not leveraging this knowledge negatively impacts the sustainable management of natural resources.

The value of GIS software and PGIS for First Nations communities brings forth a number of inquiries relating to the use of these technologies for First Nations' land protection, community development, and knowledge preservation. This chapter seeks to explore the following questions:

1. What are the benefits of GIS technologies for First Nations communities?
2. How are First Nations communities currently accessing GIS technologies, and how are they using them?
3. How can we make GIS technologies more accessible for First Nations communities?

In this chapter, I will discuss PGIS, Indigenous knowledge systems, and sustainable community design, focusing on First Nations communities in Canada. First Nations communities practice principles of ecological kinship and stewardship, which pertain to the field of the ecological humanities. These knowledge systems situate Inuit, Metis, and First Nations communities as "caretakers of Mother Earth," who "have a special relationship with the earth and all living things in it" (Assembly of First Nations n.d.). Indigenous knowledge systems are based on hundreds of years of living as part of the environment and should be recognized as an important tool for natural resource management. Naturally, Indigenous knowledge systems understand that "land use, changes in land use, and climate change are profoundly linked" (Townsend, Moola and Craig 2020). Sustainable community design, in turn, helps structure human communities to exist in harmony with the natural environment and minimize or eliminate land use practices that can harm the environment. By situating the benefits of GIS technologies as part of sustainable community design, I also discuss how arming First Nations communities with these data tools can help advance First Nations activism on preserving the land and its resources. In doing so, this chapter contributes to knowledge on these emergent tools, while also exploring their potential for producing activist knowledge.

HISTORICAL CONTEXT

To understand why some First Nations communities are currently seeking out GIS technologies, it is important to understand the history of how First Nations people have struggled to have a voice in land and natural resource management activities. In 1997, after 347 days of the trial of *Delgamuukw v. British Columbia*, the Supreme Court stipulated that "governments be obliged to consult with Aboriginal groups regarding potential development activities on their traditional lands" (Whalen 2006). This decision was the result of efforts by thirty-five Gitxsan and thirteen Wet'suwet'en hereditary chiefs, who filed their statement of claim with the British Columbia Supreme Court because British Columbia would not participate in the land claims process, despite receiving the Gitxsan declaration of claim in 1977, and instead began to allow clear-cut logging into the Gitxsan and Wet'suwet'en territory without the permission of the hereditary chiefs (Jang 2017, Smith 2004). Following the Delgamuukh precedent, in *Tsilhqot'in Nation v. British Columbia* (2014), after a five-year trial, the Supreme Court ruled that the Tsilhqot'in had claim of Aboriginal title to the 1,750 square kilometers (680 sq mi) region they had historically occupied.

Since the Delgamuukw decision, many Indigenous communities across Canada have been able to take on a more active role in natural resource management. However, this role has not come without additional and lengthy struggles. First Nations' engagement with provincial and federal judicial systems and governments has been a complex battleground for First Nations communities, where these communities have had to expend their energy and resources toward lengthy processes, and often without satisfactory results. What proved to be useful in both *Delgamuukh v. British Columbia* and *Tsilhqot'in Nation v. British Columbia* was a clear articulation of what was considered First Nations' territory. Nonetheless, there are many challenges associated with mapping Indigenous geographies. Historically, mapping has been used by governments to "establish modern territories at the expense of local claims" (Roth 2009). Throughout the years, much Indigenous territory has been lost due to settler mapping, which has mapped over Indigenous territory. One alternative is counter-mapping, or community-based mapping, where communities use the "technologies previously utilized by power holders and decision makers" to create alternative maps (Cochrane, Corbett and Keller 2014). However, this also has consequences: the increased privatization of land, loss of indigenous conceptions of space, and increased regulation by the state.

The idea of counter-mapping and PGIS is part of the reason why many First Nations leaders have "increasingly stressed the importance of developing in-house GIS expertise to reach land claims settlements and manage local resources" (Whalen 2006). Yet, not all First Nations communities want to engage in land claims processes as defined by the state, because of the above-mentioned consequences, and because they may not feel properly represented in these state processes. After all, from 1927 to 1951, Canada prohibited First Nations communities from retaining legal counsel (Lickers 2004). While this law was later repealed and some policy changed have occurred over the years, the First Nations peoples' greatest objections "concerned Canada's inherent conflict of interest in the claims process; a process in which it continued to act as both judge and party" (ibid).

HOW ARE INDIGENOUS COMMUNITIES USING GIS TECHNOLOGIES?

Today, many Indigenous communities in Canada have been able to derive benefit from GIS technologies, notwithstanding the entanglements of power at play. GIS technologies and practices present a range of opportunities for Indigenous communities, allowing the "translation" of traditional knowledge into high-tech visual maps. There have been several instances where Indigenous groups have been able to take advantage of geospatial technologies to promote their political agendas.

Some examples of this have already been discussed above, but others include the Innu Nation in Central Labrador, who had a recognized role in the forest management process due to their use of GIS at the Innu Nation Environment Office. This allowed them to devise a Forest Management Plan, an agreed upon share of harvesting permits between the Innu nation and the Newfoundland and Labrador Forestry Department (Whalen 2006). In northern British Columbia, the Gitxsan are integrating their Traditional Ecological Knowledge (TEK) with "modern ecosystem mapping techniques" (ibid). Traditional Ecological Knowledge can be defined as Indigenous systems of knowledge that are passed on intergenerationally, synonymous to Indigenous ways of life. However, it's worth noting that there are contentions around this concept for being Eurocentric; as such knowledge can "vary from Nation to Nation and from individual to individual," and "reducing this diversity to more universal definitions, it is believed, is a first step in the Eurocentric process of separating TEK from its intended context" (McGregor 2006). Beyond GIS, geomedia is benefiting Indigenous communities. SIKU is an Indigenous Knowledge Social Network and an app developed by Inuit Elders in the Arctic for a variety of purposes such as identifying thin ice, good hunting grounds, and polar bear sightings through GPS. This lends to the idea that these geospatial projects hold potential for blending Indigenous and scientific spatial knowledge in ways that can recognize and respect Indigenous spatial expressions. However, this rests on First Nations' ownership of their data and the way this data is used and the acknowledgment that Indigenous geographies should not need to be "transformed" through western technologies to be respected.

BARRIERS TO GIS IN INDIGENOUS COMMUNITIES

Not all Indigenous communities have access to GIS technologies. The reasons some communities use them, and some do not, can be quite varied. In the case of British Columbia, there are over 200 First Nations communities, the second-largest First Nations population in Canada (Statistics Canada 2019). Most of British Columbia, around 95 percent, is on unceded traditional First Nations territory, meaning that the "First Nations people never ceded or legally signed away their lands to the Crown or Canada" (Wilson 2018). Add to this, the outcome of *Tsilhqot'in Nation v. British Columbia* (2014), a case that recognized the title of the Tsilhqot'in nation to 1,750 sq. km of land in central British Columbia to use and manage the land and reap its economic benefits—a ruling that affects all "unceded" land in Canada (Lukacs 2014). These dynamics and legal landscape grants First Nations communities in British Columbia unique powers in comparison to other provinces and territories in Canada.

Since First Nations communities are a formidable population and consumer base in British Columbia, GIS technologies tend to be marketed to First Nations communities as "solutions." One solution that has been popular with First Nations communities in British Columbia is City Works, an asset management software that is built upon the ESRI ArcGIS platform. This software collects all the utility infrastructure information in a community and splits it into a dashboard that can be visualized. It has been used by BC First Nations to mark down traditional sites, often dubbed as "heritage sites." These are noted in the software, translated from elders' word of mouth, translating oral tradition into something visual and accessible, and allowing communities to collect information that is vital to their history and culture. ArcGIS solutions can be sold directly to a tribal council, and the tribal council will then disperse the software directly to the different bands underneath the council. If the community derives benefit, it catches on with other communities, expanding to other groups. As a result, the use of GIS technologies can be assumed to be more normalized in British Columbia, with many communities there already looking at data as the next frontier to decision-making.

At the same time, these programs are for-profit solutions offered by a private organization. First Nations communities may buy into these programs because they see the benefit to their day-to-day operations and long-term forecasting capabilities. This includes the importance of having a permanent record of data, especially with land-use titles and treaties as well as environmental research and sustainability in that space. We see evidence of this being the case with certain communities that use ESRI ArcGIS. However, not all communities have the financial resources necessary to buy into these programs or be part of word-of-mouth networks that encourage them to invest in GIS technologies. Furthermore, accessibility is limited by a longer procurement, which can take months before First Nations communities can move forward with the typical software. There is also a lack of comprehensive training in GIS, which results in a lack of GIS analysts that could make procurement of these technologies easier. If GIS solutions were to be offered through a government initiative, on the other hand, perhaps the procurement process could be shortened with the availability and support of GIS analysts and academics working in the public sector.

Accessibility barriers to GIS technologies create "have and have-nots" in First Nations communities, rather than being something that all First Nations communities can opt into. When GIS technologies are not accessible, they are not as normalized in an area or across communities. At the same time, many natural resource management processes, such as environmental assessments, use GIS technologies. As such, the unequal access to GIS technologies and training perpetuates the subjugation of Indigenous knowledge based on its formatting, while not providing an equal playing field for Indigenous communities in Canada who have a significant stake in resource development on their territories. While legal precedents from Supreme Court decisions, like *Delgamuukw v. British Columbia*, have granted Indigenous communities across Canada certain land rights on paper, additional work is needed to accommodate these justified rights to land and resource management within federal and provincial processes in order to decolonize climate action.

In the not-for-profit space, organizations are working to address issues in accessibility. For example, the First Nations Technology Council is an Indigenous-led organization that serves all First Nations communities across British Columbia to "ensure that Indigenous peoples have full and equitable access to the tools, training, and support to maximize the opportunities presented by technology and innovation" (First Nations Technology Council n.d.). Among their services, they provide free training on GIS and GPS technologies to individuals from Inuit, Metis, or First Nations backgrounds. These organizations are relatively new and emergent, with the First Nations Technology Council operating since 2013 (ibid). A desktop review of similar organizations finds that there are not many not-for-profit organizations operating in this space. This could be because the potential of GIS technologies are not as normalized for First Nations outside British Columbia, despite being normalized in the government.

The government uses GIS technology through its geospatial data infrastructure (referred to as CGDI), facilitated through a national program called GeoConnections to manage geospatial information assets. Mainly, it helps the government gain perspective into social, economic, and environmental issues by providing an "online network of resources that improve the sharing, use, and integration of information tied to geographic locations in Canada." Relying on the collaboration and partnership of stakeholders such as nonprofit actors, academia, the private sector, and others, it is a powerful decision-making tool that can assist Canada in risk-management protocols, such as environmental impact assessments and environmental risk assessments.

These protocols can benefit First Nations, Inuit, and Metis populations, who are especially vulnerable to the potential consequences of natural resource development activities. If they were used as a vehicle for communicating, these processes could advance current attempts at reconciliation by including Indigenous knowledge systems and grounding Indigenous self-determination and jurisdiction within the context of the natural environment, but, as I will describe below, in their current state, these protocols are exclusionary to Indigenous communities.

HOW GIS BARRIERS ARE BARRIERS IN NATURAL RESOURCE MANAGEMENT

These days, hard-data tools are becoming an important part of natural resource management and other decisions that affect the environment. However, there are fundamental inequities in how environmental assessments are carried out, including accessibility barriers relating to GIS technologies, which leads to the continuous infringement of protected Indigenous rights. As a result, Indigenous peoples can be wary and distrustful of these protocols, as it may not be inclusive of Indigenous knowledges (Arsenault et al. 2019). This is a significant hurdle to Indigenous–state relations.

Due to the profitability of natural resource extraction activities, environmental assessments can result in changes to project design, but "rarely lead to the denial of industrial activity" (ibid). Earlier, it was mentioned that land claims processes can be considered inequitable because the state is both judge and party. Likewise, without the due consultation and equal foresight of Indigenous communities in environmental risk assessments, these protocols also prioritize state interests over Indigenous interests.

This can affect the health and sustainability of Indigenous communities, who are more vulnerable to the consequences of such activities. Indigenous communities throughout Canada are negatively affected by resource extraction. During pipeline construction, even aside from the ecological harm caused by oil spills, there are many general transportation concerns, including the loss of organic matter in the soil, increasing water-off rates, increased likelihood of forest fires, temperature changes in the water, and potential harm to fish species, which are "highly sensitive to sediment release from construction" (Walden and Rozhon 2012). When oil spills happen, they can impact community drinking water. Worse still, clean-up efforts can be delayed or even halted altogether when there is a lack of communication between the private sector and First Nations. In 2016, following an oil spill in the North Saskatchewan River, two First Nations filed a multimillion-dollar lawsuit against Husky Energy, alleging that Husky failed to take accountability of the spill and take remediation steps to "mitigate, limit, or remove the adverse impact of the spill" (Eneas 2018). Now, Canada can encourage better relations between the private sector and Indigenous communities by giving more credence and platform to First Nations consultations, but this requires restructuring

GIS technologies and skills are a small part of this—a technology that is widely used by environmental professionals but still accessible to only some Indigenous communities. By instilling a technological barrier to environmental decision-making, but not providing the technological tools to participate equally, the Canadian state decentralizes Indigenous feedback and consultation from natural resource management.

As emphasized, Indigenous communities are sustainable communities. They have existed and survived in harmony with the land and environment for thousands of years and have dealt with dramatic climate changes over time, even before colonization. In "creation stories," which are passed intergenerationally, Indigenous communities learn about their relationships with and responsibilities to the environment. These stories discuss the "creation of the world and how First Peoples came to live in it." They "embody a view of how the world fits together, and how human beings should behave in it" (Canadian Museum of History n.d.).

In contrast, Western societies are somewhat separated from the environment, which can be seen as a root cause of many climate-related and environmental issues. As articulated by many environmental experts today, transformations in human-nature relationships need to happen as part of climate action. As a society, we are already seeing how unsustainable practices are harming the environment. For instance, due to unsustainable waste-management practices that prevent full recovery of materials, higher levels of plastic fragments, also called microplastics, are being found in the waters, including the binational Great Lakes Region. This is causing damage to these freshwater ecosystems, including affecting the reproductive cycle of fish populations. Its estimated that the cleanup of these plastic fragments could exceed costs of up to $400 million per year (Driedger, et al. 2015). In essence, the environmental community is struggling for solutions to present environmental concerns, while there are barriers for Indigenous communities to self-determine solutions, at least on a policy or government-to-government level. This is, by nature, a counterproductive dynamic in the environmental community.

THE VALUE OF PGIS TECHNOLOGIES

PGIS technologies are becoming increasingly relevant to the environmental community. One of the reasons that this is the case is because PGIS centers marginalized voices, which are typically left out. However, one of the main reasons for the longstanding exclusion of marginalized voices in the environmental community goes beyond GIS systems. Marginalized communities tend to be left out of environmental decision-making due to the socially constructed value of "hard" versus "soft" data in the environmental and general research community, with the exclusion of Indigenous communities in environmental assessments being a keen example of this. Hard data refers to quantitative data—hard facts that can be supported by numbers. In contrast, soft data refers to qualitative data, which can often include lived experiences. This is problematic for many reasons.

Typically, just as the first GIS systems were exclusively used by high-ranking government officials and academics, the idea of using of hard data infers possessing a technical or academic background. However, a significant body of literature describes how institutional racism can be interwoven into the very fabric of academic culture (Wilson 2018; Monture 2009; McGregor 2006). Adding to that, many marginalized and Indigenous ways of knowing are different to the knowledge and research that is valued by academic institutions. In a chapter confronting "whiteness" in the university, the late Patricia Monture, a Canadian Mohawk activist and educator, discusses the broader intellectual practices of the Haudenosaunee. She states, "Haudenosaunee knowledge systems are rational, peace-based ways of locating 'man' in society" (Monture 2009). Through her wording, these ways of thinking are based on intrinsic rationality, which "moves the thinker beyond emotion, reaction, and prejudice" (ibid). However, she sees this way of thinking as being in contrast with what is valued in the university, which she describes as a "White institution." The requirement to possess technical or academic qualifications in order for one's knowledge to be valued is not only inaccessible to some, but inherently discriminatory to racialized and Indigenous peoples.

Fortunately, this barrier or stigma against soft data is lessening, in favor of attracting a diversity of input toward climate action and environmental decision-making. Take, for instance, the Youth Climate Report, which uses a GIS map, called "Geo-Doc," to showcase over 400 self-shot documentaries produced by youth around the world, forming a "foundational micro library of [GIS data] assets" (Terry n.d.). The Geo-Doc is free and accessible, and, as an emergent tool for activist knowledge, can share "counter-knowledges" to a global audience. This can direct focus and attention to environmental issues that are experienced in a certain location or track environmental issues across nations. In the context of Indigenous communities, the Geo-Doc can potentially fill a communications gap and, better still, rather than transforming Indigenous lnowledge into hard data, allows it to remain as soft data.

Like the CityWorks platform, the Geo-Doc can be a "solution" for First Nations communities in Canada that allows them to translate oral knowledge into a technological platform. Both platforms have their value in specific situations, but the participatory format of the Geo-Doc is different because it allows all knowledge to be valued, with the principle that the more data is added, the richer the platform gets.

CONCLUSION

The question of how best to include Indigenous communities within GIS platforms is not easily answered. There are many Indigenous communities throughout Canada currently using GIS platforms in a number of ways from community development, to resource management, to knowledge preservation. These communities have historically been excluded and discriminated against, and though their experiences cannot be generalized, for many Indigenous communities, GIS platforms can be appealing as a method to participate in environmental decision-making, to adapt to a technology-based world, or to translate their knowledge systems into a format that is easily accessible to a general public audience.

However, many communities who could potentially benefit from these technologies are currently unable to do so. GIS skills training is not readily offered, and the provision of GIS solutions is still largely in the private sector's domain, which can create hurdles to procurement and access. Not all Indigenous communities throughout Canada are afforded the same jurisdictions under Canadian law, as can be demonstrated by the treaty history and current ongoing BC Treaty Process in British Columbia. At the same time, beyond issues of access, Indigenous knowledge systems are not disseminated in the same regard as Western knowledge systems. While some Indigenous Knowledge is repackaged as "Traditional Ecological Knowledge," or "TEK," this only perpetuates an unfair valuation of Indigenous knowledge as a whole. PGIS systems are shown to address some of the barriers to accessibility and also address some broader barriers of knowledge valuation. Nonetheless, reshaping GIS systems is only a small ripple in a grand scheme of Indigenous displacement.

In consideration of the deep-rooted inequalities that prevent the equal participation of Indigenous communities in environmental decision-making, it is important to reshape these platforms to be more inclusive to Indigenous populations. As a research community, how can we provide solidary for Indigenous knowledge systems? And, as an environmental community, how can we make space for those with valuable knowledge, who are restricted by a number of intersectional barriers and bureaucratic red tape? These questions are, perhaps, even more difficult to answer, but center a powerful discussion of how access and equity are often denied to valuable, Indigenous knowledge-holders and draw meaningful attention to how Whiteness and colonial mindsets function as a praxis in the dissemination of environmental knowledge.

REFERENCES

Arsenault, Rachel, Carrie Bourassa, Sybil Diver, Deborah McGregor, and Aaron Witham. "Including Indigenous Knowledge Systems in Environmental Assessments: Restructuring the Process." *Global Environmental Politics* 19, no. 3 (2019): 120–132. muse.jhu.edu/article/733985.

Assembly of First Nations. n.d. *Honouring Earth*. https://www.afn.ca/honoring-earth/.

Booth, Annie, and Norm W. Skelton. "'We Are Fighting for Ourselves:' First Nations Evaluation of British Columbia and Canadian Environmental Assessment Processes." *Journal of Environmental Assessment Policy and Management* 13, no. 3 (2011): 364–404. https://www.worldscientific.com/doi/abs/10.1142/S1464333211003936

Canadian Museum of History. n.d. *Canadian Museum of History*. https://www.historymuseum.ca/history-hall/traditional-and-creation-stories/.

Cochrane, Logan, John Corbett, and Peter Keller. *Impact of Community-Based and Participatory Mapping*. Victoria, British Columbia: University of Victoria, 2014.

Craig, W. J., T. M. Harris, and D. Weiner. *Community Participation and Geographic Information Systems*. New York, NY: Taylor & Francis, 2002.

Driedger, Alexander, H., Hans Dürr, Kristen Mitchell, and Phillippe Van Cappellen. "Plastic Debris in the Laurentian Great Lakes: A Review." *Journal of Great Lakes Research* 41, no.1 (2015): 9–19. http://dx.doi.org/10.1016/j.jglr.2014.12.020

Eneas, Bryan. *Global News*. July 24, 2018. https://globalnews.ca/news/4351120/first-nations-lawsuit-husky-oil-spill-north-saskatchewan-river/.

First Nations Technology Council. n.d. *About FNTC*. https://technologycouncil.ca/about/.

Jang, Trevor. *The Discourse*. February 7, 2017. https://thediscourse.ca/urban-nation/twenty-years-historic-delgamuukw-land-claims-case-pipeline-divides-gitxsan-nation.

Kelly, Lindsay. *First Nation Using GIS Data to Map Cultural Values*. November 22, 2012. https://www.northernontariobusiness.com/industry-news/mining/first-nation-using-gis-data-to-map-cultural-values-369087.

Lickers, Kathleen. *Looking Forward, Looking Back: Canada's Response to Land Claims*. June 2004. https://www.culturalsurvival.org/publications/cultural-survival-quarterly/looking-forward-looking-back-canadas-response-land-claims.

Lukacs, Martin. "The Indigenous Land Rights Ruling That Could transform Canada." *The Guardian*. October 21, 2014. https://www.theguardian.com/environment/true-north/2014/oct/21/the-indigenous-land-rights-ruling-that-could-transform-canada.

McGregor, Deborah. "Traditional Ecological Knowledge." *Ideas: the Arts and Science Review* (2006). http://www.silvafor.org/assets/silva/PDF/DebMcGregor.pdf

Monture, Patricia. "'Doing Academia Differently:' Confronting 'Whiteness' in the University." In *Racism in the Canadian University: Demanding Social Justice, Inclusion, and Wquity*, by Henry Frances and Carol Tator, 85–115. Toronto, ON: University of Toronto Press, 2009.

Rambaldi, Giacomo, Peter A. Kwaku Kyem, Mike McCall, and Daniel Weiner. "Participatory Spatial Information Management and Communication in Developing Countries." *Electronic Journal of Information Systems in Developing Countries* 25 no. 1 (2005): 1–9. http://dx.doi.org/10.1002/j.1681-4835.2006.tb00162.x

Roth, Robin. "The Challenges of Mapping Complex Indigenous Spatiality: From Abstract Space to Dwelling Space." *Cultural Geographies* 16, no. 2 (2009): 207–227. https://doi.org/10.1177%2F1474474008101517

Smith, Jane M. "Placing Gitxsan Stories in Text: Returning the Feathers, Guuxs Mak'am Mik'aax." *The University of British Columbia.* September 2004. https://open.library.ubc.ca/cIRcle/collections/ubctheses/831/items/1.0054675.

Statistics Canada. *First Nations Population by Provinces and Territories, Canada, 2016.* February 7, 2019. https://www150.statcan.gc.ca/n1/daily-quotidien/171025/cg-a002-eng.htm.

Sutton, Jeff. *A Geographic Information System (GIS) as a Tool for First Nations Housing Management, Planning, and Safety.* Ottawa, ON: Canada Mortgage and Housing Corporation, 2006.

Terry, Mark. n.d. *The Geo-Doc: Enhancing Environmental Education through Geomedia.* Accessed February 20, 2021. https://www.asle.org/features/the-geo-doc-enhancing-environmental-education-through-geomedia/.

Townsend, Justine, Faisal Moola, and Mary-Kate Craig. *Indigenous Peoples Are Critical to Nature-Based Solutions to Climate Change.* October 26, 2020. https://ipolitics.ca/2020/10/26/indigenous-peoples-are-critical-to-nature-based-solutions-to-climate-change/.

Walden, Zoey, and Jon Rozhon. *Oil Spills and First Nations: Exploring Environmental and Land Issues Surrounding the Northern Gateway Pipeline.* Canadian Energy Research Institute, 2012.

Whalen, Jake. "Distributed GIS Solutions for Aboriginal Resource Management." *The Canadian Journal of Native Studies* 25, no. 1 (2006): 139–53. http://www3.brandonu.ca/cjns/25.2/default.htm

Wilson, Kory. "Pulling Together: A Guide for Indigenization of Post-Secondary Institutions. A Professional Learning Series." *BC Campus Open Publishing*, 2018. https://opentextbc.ca/indigenizationfoundations/

Chapter Four

The Use of GIS by Indigenous Peoples in Charting Culture, Claims, and Country

Jigme Lhamo Tsering

Historically, maps have been used to visually represent significant geographical features of the land and social demarcations that designate boundaries to conquered and claimed areas. Over time, maps have become digitized contributing to geographic information systems (GIS) that are used to visualize, analyze, and store large amounts of data of geographic locations around the globe.

GIS projects can be used as a tool to challenge historically cemented colonial legacies that fail to accurately represent Indigenous lands and their cultural significance. Indigenous communities around the world are using GIS practices like biomapping to reassert their right to their territory, as they believe they are better suited to make governing decisions for their ancestral lands than other non-Indigenous parties. Indigenous knowledge is a complex system that involves a direct and sacred relationship with the Earth and offers a unique perspective on geographical caretaking.

With the ability to read weather patterns, share culturally enriched creation stories, and possess a better understanding of the environment and how to maintain it, any community can learn how to respect and prolong the health of our planet and the natural resources it provides (Chambers et al., 2004). To Indigenous people, their knowledge, voice, and geographical representation have had a big impact on their ability to pass on their culture, which is currently being threatened by modern industrial developments. According to the United Nations,

> Indigenous peoples are inheritors and practitioners of unique cultures and ways of relating to people and the environment. They have retained social, cultural, economic, and political characteristics that are distinct from those of the dominant societies in which they live. Despite their cultural differences, Indigenous peoples from around the world share common problems related to the protection of their rights as distinct peoples. (United Nations n.d.)

One of these common problems is that First Nations Peoples communities around the world are situated within international nation-states with varying political ideologies operating within Westphalian sovereignty. Today, the concept that each state has exclusive authority over its territory is being challenged through participatory Aboriginal-led GIS projects (Shadian 2014). In *The Red Deal: Indigenous Action to Save Our Earth*, The Red Nation states:

> Our sovereignty—our very being—cannot be separated from the health and well-being of the land. Like all societies and civilizations, our relationality, and the values upon which these practices are based, is what makes us who we are. Our political orders and systems of governance only matter if we caretake the land; they are intelligible through our relationship with the land—literally. We do not want to endure any longer; we want to thrive. (The Red Nation, 2021, 56)

This quote is a sharp nudge at the human rights and environmental exploitations through the colonial construct of capitalism, and a declaration that Indigenous knowledge and wisdom will be an effective tool in dismantling this framework. This statement also conveys the intricate and direct relationship between Indigenous land and peoples. The relationship between the two is represented through cultural practices that value land and sustainability for all with one statement: " . . . ensure that countries in the Global South will be able to develop sustainably and guarantee sustainable livelihoods for their citizens" (The Red Nation 2021, 36). The conservation of environmental habitats, land, and Indigenous spaces contributes directly to the continuation of Indigenous culture. The use of participatory GIS to document and conserve delicate ecosystems has charted a new way for Aboriginal peoples to speak to authorities of industrial and governmental power on the matters of culture, claim, and country.

GIS maps are important tools that are subject to their user's intent; they can maintain colonial influence or build the framework for Indigenous resilience and resistance. Focusing on the digitization of Indigenous knowledge, this chapter will explore how geospatial data is empowering a modernized approach for the representation, protection, and advocacy of Indigenous peoples and places.

The three sections within the chapter will look at how GIS has been used to connect Indigenous and non-Indigenous stakeholders, to educate and influence change, and how it has evolved to introduce accessible frameworks for inaccessible communities. Note that throughout this chapter, nations and states will be referred to in their Indigenous spelling with the Anglican name in brackets. Recognizing land using the Indigenous names for spaces is also an avenue for those of us that benefit as settlers to engage in cultural preservation.

BRIDGING BETWEEN INDIGENOUS AND NON-INDIGENOUS STAKEHOLDERS

GIS has made effective communication possible between Indigenous and non-Indigenous stakeholders using observable data to support Indigenous arguments and perspectives. For example, across Kanata (Canada), many Indigenous communities share the cultural value of respect for the environment. They believe that as one of the newer species on Earth, people can learn from the animals, places, and environmental patterns that have existed long before them to guide them to live healthy, responsible, and sustainable lives. This deep understanding of the space and matter that preexist them are evident in the line "the plants, rocks, animals, and hydrothermal vents on the seafloor to the arid mountains of colonial Chile, there are countless communities found throughout the Earth, not all of them human, and there are countless relationships that have been gifted to and cared for between these communities" (The Red Nation 2021, 34). These beliefs are supported by centuries of sustainable living practices and continue to be strengthened by their direct relationship to their environment. In contrast, it differs greatly from the extractivist and capitalistic ideals of neoliberalism, the ideological framework that policymakers and industries often function within.

This section will look at examples of GIS use by Indigenous peoples as they work to communicate their intricate connection with the land to non-Indigenous stakeholders. The use of GIS and spatial data for and by Indigenous communities began in North America about five decades ago. One of the first rudimentary cases of Indigenous communities using GIS mapping methodologies was *The Inuit Land Use and Occupancy Project* by Milton Freeman in 1976. The project was proposed by the Inuit Tapirisat of Canada to the Minister of Indian and Northern Affairs and covered thirty-three Inuit communities across the Northwest territories. It documented their perspective on past and present resource-gathering practices.

The purpose of this project was to propose legal legitimization of the Inuit right to land use and occupancy across the Arctic region and as a tool for negotiating land claim agreements that would protect the environmental integrity of their traditional lands. The latter was out of concern for the increasingly numerous and poorly regulated nonrenewable energy resource developments causing significant pollution in their hunting, fishing, and gathering grounds (Freeman 2011). Subsequently, this initiative recorded substantial data on Inuit connections to the land from the communities' perspectives including history, the names of places, and livelihood techniques, among other cultural data. Data was collected using a blend of map biographies and narrative recordings of land-use changes throughout chronological time (Freeman 2011, 22). The individual map biographies were then amalgamated to create detailed digitized representations of community land-use history. *The Inuit Land Use and Occupancy Project* was initiated with the goal of creating information and evidence that could be understood by non-Indigenous policymakers. With the support of researchers like Freeman, the project was successful in providing the base requirements to enter negotiations for the land claim. Freeman also states that *The Inuit Land Use and Occupancy Project* was wielded during the negotiations, with the project contributing to an agreed upon boundary for Inuit lands, that only had minor differences from the one that was proposed by the Nunavut Constitutional Forum (Freeman 2011, 27). The use of GIS has developed since this first example into an advanced system that is capable of documenting, analyzing, and communicating:

> land use and occupancy; sacred sites; oral histories and place names; environmental ethics; subsistence hunting, gathering, fishing, and trapping activities; intimate knowledge of fish, wildlife, and plants; travel paths and portages; historical migration; family and kinship organization; patterns of harvest sharing and consumption; seasonal cycles and variations; sea ice and permafrost; landscape changes and lake processes; and cosmological and spiritual knowledge. (McGetrick, Bubela, and Hik 2015, 177)

The ability to incorporate this information into a simplified geodata report allows for non-Indigenous stakeholders to better comprehend Indigenous interests. This use of GIS by the Inuit communities is an example for other Indigenous communities to use similar methodologies in reasserting their right to self-determination.

In more recent years, the Mbendjele community of the Congo Basin Rainforest, who are nonliterate and experiencing threats to their ancestral forests due to deforestation and poaching, also utilized GIS and non-Indigenous support in sustainable management of their forests (ArcNews 2016). The Mbendjele initiated the project, supported by the ExCiteS team, to document trees that were of cultural value to the community for medicinal or religious purposes. This resulted in the creation of Sapelli; the software's website defines it as "an open-source project that facilitates data collection across language or literacy barriers through highly configurable icon-driven user interfaces" (Sapelli 2021). Sapelli was useful in overcoming a literacy barrier and also the lack of stable cell phone service in the Congo rainforest; the software was designed to occasionally check for connectivity and transfer recorded data to a GIS server when most appropriate. The geospatial documentation through this project was important when engaging with the local logging company, Congolaise Industrielle des Bois, who then made note of these important trees and removed them from their logging plans (ArcNews 2016). GIS is a proven tool for direct communication between Indigenous changemakers and non-Indigenous decision-makers, be it those of government or industry authorities.

GEOMEDIA TO EDUCATE AND INFLUENCE SOCIAL CHANGE

The marriage between documentary filmmaking and other forms of media alongside GIS is ushering in an innovative new avenue for dynamic spatial data documentation. Geomedia in this sense is producing an effective way for Indigenous nations and specifically Indigenous youth to take a direct role in presenting their case in a way that policymakers can easily comprehend (Terry 2020). Both the Inuit Tapisirat and the Mbendjele saw a level of success in their use of GIS to legitimize their right to territory and create change in land-governing norms in their respective regions. There exists a larger conversation on the transnational environmental impact of decisions made within Westphalian sovereignty that excludes Indigenous representation, particularly in the Arctic, as climate change creates easier access to the circumpolar region and its valuable resources the geopolitical conversation of Arctic ownership reignited. The circumpolar Arctic is sparsely populated with less than 10 million people over 17 million square kilometers (McGetrick et al. 2015). Nevertheless, even the most isolated Indigenous peoples could be impacted by international decisions made in faraway cities, especially through the extraction and export of Arctic commodities around the world. On this, Jonathan Greenberg states:

> No community or environment in the Arctic remained untouched and unchanged by the larger historical forces, ideologies, and economic engines that have shaped the international system as it evolved since the beginning of the European nation-state system. (Greenberg 2009, 1315)

This statement conveys that despite the Arctic being in a geographically expansive and isolated region, it is still subject to the side effects of international neoliberalist decision-making. This is further corroborated by the

ArcGIS StoryMap that explains the negative health impact associated with extractivist oil and resource drilling in the Arctic region (Herrera-Bevan 2020). This approach utilized case reports rather than film to showcase the range of communities facing similar struggles. The use of geomedia as a communication tool removes the need for extensive scientific understanding, information can be conveyed in a simple and moving manner through film.

The Youth Climate Report (YCR) GIS Project is a prime example of a geomedia platform that is generally accessible, amalgamating place-based reports into an accessible and dynamic data atlas that is used to influence global discussions (Terry 2020). Using platforms like the YCR allows for better engagement with a growing generation of changemakers, Indigenous youth, in their efforts to voice their concerns for their communities. An example of this is *Happening to Us*, a documentary that depicts the effects of climate change on the regional environment and infrastructure and the shifting relationship that the Tuktoyaktuk community has with their land (Gruben et al. 2019). The film opens with eroding coastlines and the makeshift barriers that the community has created to protect their homes from flooding. In one scene, one of the youth filmmakers interviews an elder who orates how traditional practices that rely on the ice have been affected due to climate changes. The film was presented at the 25th Conference of the Parties (COP25) to policymakers that influence global environmental governance decisions. The potential to assist dialogue and collaborative processes across the transnational Arctic region is evident; however, it would be strengthened with more educational initiatives. *Happening to Us* is the result of a film workshop, and incorporating geomedia education into schooling could encourage more youth to speak their truths to power. In addition, broadening the digital infrastructural supports systems in Indigenous communities across the Arctic would produce upward mobility and self-determination in Indigenous communities (Koncan 2014; Pierre 2019). This could also contribute to Indigenous-led initiatives in exploring avenues to return the responsibility of stewardship and land governance decision-making back to these communities. As seen in the case of the Inuit, and in repeatedly marginalized Indigenous communities around the world, the current approach toward land governance can be exclusionary toward these frontline communities. While there are efforts to create protective legislation for the Arctic region, there is a lack in the practical aspect of implementing them during inherently detrimental practices of resource extraction. The only way to mitigate this would be to put the interests of Inuit communities and the Arctic environment above those of economic interest.

ACCESSIBLE FRAMEWORKS FOR INACCESSIBLE COMMUNITIES

The conditions that influence access to GIS use are geographic location, access to resources, and political ideology. In *The Geo-Doc: Geomedia, Documentary Film, and Social Change,* Mark Terry writes about a "digital divide" that affects communities that have limited access to digital technology and resources for wireless networking (Terry 2020, 136). The inaccessibility to resources and the training necessary to be able to use and contribute to geodata creation and documentation is another factor that inhibits engagement of GIS use. An example of this would be the lack of access to stable wireless internet, electricity, and/or information communication technologies across the approximate 30 million nomadic and Indigenous peoples across Africa (Koncan 2014). While the distribution of these technologies is lacking in these regions, this can be remedied through the introduction of initiatives and programming. An example of success in the African region is the aforementioned Mbendjele Indigenous group in mapping their tribal lands. Despite difficulties due to their geographic location and experiencing a lack of access to digital technology, electricity, and cell phone signals, they were able to work with non-Indigenous stakeholders in overcoming these barriers through specific software, training, and creative technology (ArcNews 2016). Using intuitive and simple mapping technology, the Mbendjele were able to preserve parts of the land that were especially significant to their community, including trees with medicinal and religious importance.

Limitations associated with the first two conditions can be overcome through engaging in participatory Indigenous GIS use and financial support from interested parties. Communities that are geographically isolated or otherwise difficult to access, like in the case of the Inuit in the Arctic region, have benefitted from GIS tools to foster better understanding and influence policy decisions made by non-Indigenous stakeholders (Freeman 2011; McGetrick et al. 2015).

While utilizing GIS as a communication tool has proven successful for stakeholders, another barrier to its use is the cost of the equipment and maintenance needed for collecting and storing geodata. Indigenous communities located in the global north were better able to afford the accompanying expenses with the funding provided by the government through resource sales with industries (Chapin et al. 2005, 622). This can be seen in the case of Kanadario (Ontario) where six First Nations of the Iroquois community—the Mohawk, Oneida, Tuscarora, Seneca, Cayuga, and Onondaga communities—came together to conduct their own research into their ancestral land claims using GIS: the Haldimand Treaty. The provincial and federal governments did not provide direct funding for these initiatives so the fruition of the Six Nations was an indication of the sheer willpower of these communities. Over multiple decades the communities found overwhelming evidence, through historical documents, pointing to their ownership of 950,000 acres of land granted by the 1784 Haldimand Treaty, including parts of Kitchener-Waterloo and Cambridge (Six Nations Lands & Resource Department 2010; ESRI 2021).

The third barrier, political ideology, is slightly more intricate to maneuver. An important aspect to consider when examining the use of GIS by Indigenous peoples is whether there is a political will that supports it. Nations that may not want to relinquish their colonial rule over Indigenous land may try to suppress access to resources that threaten the government's Westphalian sovereignty. The political climate can have a heavy influence on Indigenous nations' access to tools, including open source geodata like the YCR. This is apparent when examining the map of the YCR, which boasts over 500 videos on place-based environmental research from its youth contributors, but there are certain areas of the atlas that remain bare, including Russia and China (Youth Climate Report 2021). This could be due to the lack of free access to open source GIS, it could be the extreme consequences associated with resistance against the government, or it could be a blend of the two. While the lack of access to digital resources in remote communities, as seen in the case of the Mbendjele, can be remedied through the collaboration of Indigenous and non-Indigenous actors, the task of bridging digital access grows significantly more difficult in nations that have different governing systems than those that value democracy (ArcNews 2016). But it is not impossible. A specific case is the cohesive efforts of Tibetan Indigenous people inside Tibet and those that are displaced outside of Tibet. With the use of systems like the Environmental Justice Atlas, resisting Indigenous Tibetan nomadic and pastoralist communities endeavor to protect the integrity of their environment from the threat of industrial advancements, through digitization and documentation (EJAtlas 2021). This is not unique to the Tibetan people and serves as preliminary documentation of the significance of land for cultural preservation and practices. The trajectory of GIS-based evidence for engaging in conversations for more sustainable and Indigenous governed land use. This could be the beginning of a shift in these more contested regions. Overcoming the roadblocks that indigenous communities face in communicating their concerns and advocating for their rights is becoming increasingly possible through the technological advancements of GIS and the increasingly creative ways they are being used.

CONCLUSION

The topic of this chapter has been to perform a critical analysis on the way GIS has shaped the way Indigenous communities chart culture, claim, and country. These communities are often on the frontlines of climate change and are among the first to feel the direct impacts of pollution and other environmental justice concerns. Consequently, using participatory GIS methodologies in documenting Indigenous occupancy, land use, and place-based knowledge have resulted in success (Freeman 2011; ArcNews 2016). Indigenous lands and cultures are intricately entwined with one another, although GIS projects have paved the way for effective communication with non-Indigenous leaders, where inaccessibility is an issue. This chapter has examined multiple case studies of GIS use among Indigenous communities, and a common theme across all of them has been a deep concern for the environment. This is a rapidly growing concern as the effects of climate change continue to increase with natural disasters. Extreme heatwaves, forest fires, and flooding are global concerns that extend beyond Indigenous communities. Increasing accessibility by providing Indigenous populations with access to necessary resources, and political support can not only propagate Indigenous advocacy and self-determination but also assist in protective environmental practices.

Future examinations of Indigenous GIS use could benefit politically marginalized indigenous groups that are not recognized as deserving members of decision-making communities. Clarifying who decides the "Indigenous status" of a group of people, for example, in the case of China, could allow for fairer and more equitable policy-making in those regions. For example, Tibetans, Southern Mongolians, and Uyghur groups are considered ethnic minorities of the Chinese state rather than being rightfully recognized as Indigenous groups (Hinden 2021). This label prohibits the participation of these communities in international conversations pertaining to Indigenous governance. Silencing Indigenous peoples not only prevents them from effectively passing cultural knowledge from one generation to the next but also prevents the global community from benefiting from their understanding of our world.

REFERENCES

ArcNews. "Mapping Indigenous Territories in Africa." ESRI, 2016. Last modified 2016. https://www.esri.com/about/newsroom/arcnews/mapping-indigenous-territories-in-africa/

Chambers, Kimberlee, Corbet, Jonathan., C Keller., and Colin Wood. "Indigenous Knowledge, Mapping, and GIS: A Diffusion of Innovation Perspective." *Cartographica* 39, no. 3 (2004): 19–31.

Chapin, Mac, Lamb, Zachary, and Threlkeld, Bill. "Mapping Indigenous Lands." *Annual Review of Anthropology* 34, no. 1 (2005): 619–38. https://doi.org/10.1146/annurev.anthro.34.081804.120429

EJAtlas. "EJAtlas—Global Atlas of Environmental Justice." 2021. https://ejatlas.org/.

ESRI. "Six Nations Adapts Traditional Beliefs to New Technology with GIS." 2021. https://www.esri.com/news/arcnews/spring00articles/sixnations.html

Freeman, Milton M. R. "Inuit Land Use and Occupancy Project: a Report." Minister of Supply and Services Canada, 1976.

Freeman, Milton M. R. 2011. "Looking Back—and Looking Ahead—35 Years after the Inuit Land Use and Occupancy Project." *The Canadian Geographer* 55, no. 1 (2011): 20–31. https://doi.org/10.1111/j.1541-0064.2010.00341

Greenberg, Jonathan D. *The Arctic in World Environmental History*, 42 Vanderbilt Law Review 1307 (2021). Link: HYPERLINK "https://nam02.safelinks.protection.outlook.com/?url=https%3A%2F%2Fscholarship.law.vanderbilt.edu%2Fvjtl%2Fvol42%2Fiss4%2F8&data=05%7C01%7Camanasterli%40rowman.com%7Cad06e4a29b5a4b7291b508da8a9da72d%7C8fdc2247c6bb43e686abb8ce3c37e4bf%7C0%7C0%7C637974706972450686%7CUnknown%7CTWFpbGZsb3d8eyJWIjoiMC4wLjAwMDAiLCJQIjoiV2luMzIiLCJBTiI6Ik1haWwiLCJXVCI6Mn0%3D%7C3000%7C%7C%7C&sdata=SHjP7Iqg0ZNg60C3xa1ILq0D8v%2F3tCWdFl7hBdFH6ZI%3D&reserved=0" https://scholarship.law.vanderbilt.edu/vjtl/vol42/iss4/8.

Gruben, Kikoak, B., C. Kuptana, N. Kuptana, E. Lugt, G. Nogasak, and D. Tedjuk. *Happening to Us*. 2019. Retrieved from: https://www.youtube.com/watch?v=nE3uxUjca3.

Herrera-Bevan, Ezekiel. "Indigenous Peoples, Climate Change, and Concurrent Risks." Virginia Wesleyan University, 2020. Last modified December 2014. https://storymaps.arcgis.com/stories/59a7360d9c9b466686ba9198d35267ac.

Hinden, Adam. "Does China Have Indigenous Peoples?" Center for World Indigenous Studies, 2021. Last modified July 2019. https://www.cwis.org/2021/07/does-china-have-indigenous-peoples/.

Koncan, Alfonz. "A Literature Survey of the Global Digital Divide and Indigenous People." Athabasca University, 2014.

McGetrick, Jennifer, Tania Bubela, and David S. Hik. "Circumpolar Stakeholder Perspectives on Geographic Information Systems for Communicating the Health Impacts of Development." *Environmental Science & Policy* 54 (2015): 176–84. https://doi.org/10.1016/j.envsci.2015.07.00

Pierre, Dennis. "How Closing the Digital Divide Can Help Indigenous People Assert Self-Determination." First Nations Technology Council. Last modified October 2019. https://technologycouncil.ca/2019/10/28/how-closing-the-digital-divide-can-help-indigenous-people-assert-self-determination/.

Sapelli. 2021. Retrieved from Sapelli.org.

Shadian, Jessica. "Building an Arctic Regime." In *The Politics of Arctic Sovereignty*. Routledge, 2014. https://doi.org/10.4324/9781315851419-17.

Six Nations Lands & Resources Department. "Land Rights: A Global Solution for the Six Nations of the Grand River." 2010. https://www.tidridge.com/uploads/3/8/4/1/3841927/snglobalsolutions-web.pdf.

Terry, Mark. "The Geo-Doc: Geomedia, Documentary Film, and Social Change." Switzerland: Palgrave Macmillan, 2020.

The Red Nation. "The Red Deal: Indigenous Action to Save Our Earth." Common Notions, 2021.

United Nations. "Indigenous Peoples at the United Nations." Retrieved from: https://www.un.org/development/desa/indigenouspeoples/about-us.html

Youth Climate Report. 2021. http://youthclimatereport.org/

Chapter Five

Ecofeminist Visualization

Reading GIS as a Bridge to Gendered Water Management in India

Pamela Carralero

In the early 2000s, feminist researchers and critics sought to nuance critical discourse on geographic information systems (GIS) to establish a new politics of GIS vision uniquely centered on embodiment and gender difference. A "feminist visualization" (Haraway 1991) of GIS space consequently emerged that offered an alternative to the scientific positivism that had largely characterized the traditional geographic gaze. Coined by Donna Haraway and expanded upon by feminist geographer Mei-Po Kwan, the term feminist visualization describes "a possible critical GIS practice in feminist research" (Kwan 2002) that works to open GIS to feminist frameworks by recalibrating the very value of data. For example, in her 2002 essay, "Feminist Visualization," Kwan urges researchers to carefully consider how space and data reflect and affect women's movements and agency, thereby prompting consideration of how GIS visualizations contest or perpetuate forms of gender oppression. More specifically, then, feminist visualization registers the heterogeneities and specificities of the lived bodies traced in GIS practices while coupling visual and material theories to analyze the intertwined and gendered dynamics of data, space, and the female body.

To read feminist visualization as a call for gender-inclusivity in GIS theory and practice raises questions concerning feminist visualization's own geographic heritage and its associated limitations. As will be seen, feminist visualization is a specific response to Western visual regimes and Western gender discourse, which inadvertently signals the nature of its own selective optic and applicability. Its focus on rethinking the symbolic interpretation of women—that is, how women are represented—within the GIS map falls short as a mode of feminist empowerment when placed in nonwestern feminist research contexts. This is especially true when the challenges facing nonwestern women's movements and issues include not only breaking from limiting sociocultural representations and gender norms but also engaging in ongoing struggles for women's political representation and basic rights. While an acknowledgment of the differences between worldwide feminisms is an established issue in women's studies, feminist visualization's Western roots signal that such an acknowledgment has yet to be incorporated into feminist GIS. A nonwestern feminist GIS is nonexistent despite the growing importance of geospatial technologies to anticipating and mitigating gender vulnerabilities in the Global South, especially in relation to environmental impacts on gender-specific land use and water management.

In this chapter, I propose an ecocritical GIS framework that can support further action toward women's water equity in India. I offer feminist visualization a sister term, ecofeminist visualization, to articulate a mode of visual analysis that couples the data-space-body nexus with socioculturally specific discussions of women's oppression and rights in environmental contexts. Ecofeminist visualization implicates feminist GIS and its mode of seeing with the intense focus on materiality found in the field of material feminism. It describes a whorled visual-material lens that fashions new perspectives on national space and sociopolitics within ecological crises. Enacting ecofeminist visualization as a mode of GIS map reading and data interpretation that can affect women's everyday lives, I apply ecofeminist visualization to the materiality of India's water politics to redress the voiding of rural Hindu women's water work in water management contexts. Despite the fact that India's constitution ensures equality for all citizens before the law, women continue to be excluded from decision-making realms, their political subjecthood often tokenistic in the face of deeply entrenched differential citizenships of caste, class, and gender. While rural Hindu women "play important roles in [India's] national economy as small-scale farmers and irrigators" (Lahiri-Dutt 2009), they are discounted as formal water users by socially embedded systems of male privilege that remain blind to women's irrigation needs and lead water professionals to exercise selective perspectivism when choosing how to include women in water management planning and practice (Lahiri-Dutt 2009).

The goal of ecofeminist visualization is to urge gender-inclusive water management by making women's water work irrefutable to decisionmakers as crucial parts of India's water future. In the first section of this essay, I define ecofeminist visualization in more detail and introduce GIS data's capacity to affect, as opposed to simply represent, bodily materialities before discussing the structured violence of gender-water inequities in rural Hindu

India. The second section evinces this violence at work in a GIS blueprint that conceptualizes an ambitious water infrastructure project issued by India's water-scarce state of Gujarat. Since I hold that women's legitimation as water users is dependent on cultural shifts that change how GIS space is visualized and understood, the final section puts ecofeminist visualization into action to present an enriched mode of GIS reading that deepens visualizations of space to further gender-just water practices.

ECOFEMINIST VISUALIZATION

Ecofeminist visualization transports material politics into GIS practice to investigate the multibodied sociospatiality of GIS maps (Barry 2013). In other words, it pays attention to how human and nonhuman bodies are characterized and geospatially arranged by both dominant and alternative modes of seeing—and to what end. Ecofeminist visualization thus extends and deviates from the traditional stanchions of feminist GIS discourse, including 1990s feminist critiques of a GIS view-from-nowhere that asserts knowledge as unconnected to personal or collective experience and produces objective maps of space and bodies' spatiality (Kwan 2002, Bosak et al. 2005). It additionally diverges from feminist GIS mappings in the early 2000s that traced and analyzed "the geometry of women's life-paths"—such as the way women of different demographics move through a city—"and processes of identity formation" (Kwan 2002, Pavlovskaya and St. Marin 2007). In fact, as a way of reading GIS maps, ecofeminist visualization inverts traditional GIS methodology. While the feminist GIS researcher traditionally uses data to determine how gender representations and bodies' movements result from sociocultural organizations of space, the ecofeminist viewer analyzes how such organizations visualize certain kinds of data at the expense of others. Rather than extract a discussion of gender politics through the limited or free movement or placement of gendered bodies, the ecofeminist viewer pays attention to space as an engineer of a nation-state's gender politics. She not only considers who (genders, classes, castes, religious groups, cultural groups) and what (practices, traditions, infrastructures, environments) are made visible through GIS data but also focuses on who and what are left unmapped and invisible, omitted from view by a politics of statistical and imagistic forefronting that constructs exclusive cartographic narratives.

The importance of unmapped data to gender and material politics has yet to be fully considered by feminism as a useful form of power, which I define here in accordance with Patrick Joyce and Tony Bennett as "a condition of action" made manifest "in the practices through which it is performed and exercised" (Joyce and Bennett 2010). Perhaps one reason for this lacuna is that the relationship between feminist GIS and data remains strained despite the now-accepted truism that "data never do speak for themselves" (Keller 1992) but rather presuppose cultural and gender interpretation. In fact, feminist GIS still grapples with the ways in which mapped Cartesian coordinates seemingly annex discussions of women's lives and bodies. As Kwan points out, the "abstract geometry of points and lines cannot reflect many significant aspects of women's experiences" (Kwan 2002), making it "difficult to imagine a GIS production that can do justice to the contribution of feminist theories of corporeality and subject formation" (Kwan 2002). She goes on to suggest a possible feminist "appropriation" (Kwan 2002) of GIS methodology by reimagining coordinates as "*body inscriptions*—inscriptions of oppressive power relations on women's everyday spatiality and inscriptions of gendered spatiality in space-time" (Kwan 2002). Static coordinates thus become theoretically transformed into expressions of embodied subjectivity, giving GIS visualizations heightened context and meaning for the feminist GIS user.

However, a feminist appropriation of GIS methods also limits possibilities to imbricate critical GIS with women's studies as a field inclusive of transnational feminisms. At the heart of this problematic is the entanglement of feminist GIS in a Western geospatial hermeneutic that, in its effort to combat gender-blind geographies, suppresses questions and problems of difference amidst global feminisms. Kwan's use of *appropriation* to make quantitative geography "more compatible" (McLafferty 2005; Mattingly and Falconer Al-Hindi 1995) with feminist research and practice gestures to this unconscious foreclosing by limiting feminist GIS to a struggle against women's symbolic construction. Here, Kwan walks in lockstep with early Western feminist work dedicated to both struggling against the primacy of scientific positivism and refashioning how women were thought of, visualized, described, and otherwise socially represented. Yet what this lockstep means for GIS is that while female embodiment and subjectivity are mapped, they are only mapped as inscriptions of oppressive power relations generated out of Western gender hermeneutics. The fact that the female body appears on the map as a counter to the positivist objectivity of the quantitative inadvertently freezes her into a product of oppression, an antithesis responding to the thesis of scientific positivism rather than something in and of herself. Feminist GIS thus seemingly continues to struggle with the notion of representation, which limits its possibilities as a technology of feminist power.

Feminist epistemology has historically upheld positivism as a predominant source of women's oppression, since its claim to fully understand social and natural systems through empirical observation abnegates alternative, embodied ways of seeing and knowing. Indian feminist scholar Uma Narayan, however, cautions women's studies scholars against generalizing the quantitative as feminism's foremost challenge. As stated, the twentieth-century preoccupations with positivism that inform and percolate within Kwan's discussion arise in the context of Western-born feminisms' concerns with discursive constitutions of gender and women's social positioning. These, as Narayan points out, do not necessarily reflect non-Western women's experiences of oppression, whose conditioning environment for feminist struggles has been marked by multiple, intersecting axes of social, economic, and political inequality that not only discursively figure women but also directly restrict their movements, freedom, and human rights (Kannabiran 2010). Narayan therefore warns against generalizing feminist epistemologies, challenges, and goals into a homogenous enterprise when she writes that, in comparison to positivist frameworks, "nonpositivist frameworks are not, by virtue of that bare qualification, any more worthy of our tolerance. . . . We must fight not frameworks that assert the separation of fact and value but frameworks that are pervaded by values to which we, as feminists, find ourselves opposed" (Narayan 2004). Narayan's call for Western feminism to nuance its approach to the quantitative and disengage from binary-oriented crusades notably unfurls more critically rigorous and transversal angles from which to address the forms and contexts of women's issues. This shift in focus opens opportunities to think data and the quantitative heterogeneously as a political site beyond Western laboratory life and language, one that is both immanent in and constructive of women's differences and everyday realities.

At the crux of Narayan's argument is a GIS-related injunction that ecofeminist visualization takes up in force: The task of critical feminisms currently engaged with data practices is not in appropriating data discourse but in forging lines of inquiry into how mapped data structure relations visualizes gender equity and women's empowerment. Here, data become key as a force that can produce fresh understandings of material and social organizations, reframe sociopolitical spaces, and shape processes of social transformation, thereby taking on an affective capacity as a force or entity that can influence thought and action. While in traditional feminist geography GIS maps bodies in space, ecofeminist visualization presents data as that which maps space itself, imbuing data with a political purchase in its own right and, by extension, calling attention to the material politics that bring bodies of data and GIS mapping into being. As Bruno Latour states in reference to oxygen-producing bacterium and their importance to earthly life: "Bacteria are not *in* the frame, they *make* the frame" (Schultz and Latour 2020). Similarly, data, for the material relations that they point to, are not in the GIS map, they make the GIS map. In this way, ecofeminist visualization sits in relation to recent feminist materialist and ecocritical concepts such as Stacy Alaimo's transcorporeality, Karen Barad's intra-action, and Kathryn Yusoff's billion "Black Anthropocenes," all which reconceptualize space as that which is construed by the agency of life forms (Alaimo 2008, Barad 2007, Yusoff 2018).

Ultimately, ecofeminist visualization conceptualizes GIS data as a condition of feminist action insofar as GIS practitioners create data structures that support a stage for feminist engagement. In the context of India's water management, such disclosure seeks to reconfigure the existing, oppressive spatiotemporality of hydropower to bring equal opportunity laws into action and mitigate women's vulnerability to the climate-changing future. In Hindu communities, the social politics surrounding the use of rural water infrastructure is deeply entrenched in a context of gendered boundaries, with women's water work tethered to specific domestic-oriented water sources, including communal spigots and wells. Disparities arise in moments when water needs urge women to cross sociocultural boundaries and use canals and water pipes reserved for the imagined, abled, and development-enabling citizen who, as Kuntala Lahiri-Dutt reflects, "is almost always male—a farmer needing irrigation, making decisions with regard to water use, and belonging to that elusive public arena that is devoid of the presence of and deliberations on women's needs" (Lahari-Dutt 2009). During monsoon season in Gujarat's region of Sauarashtra, for example, water from irrigation reservoirs is reallocated "from low-valued agricultural uses to high-valued drinking and industrial use by urban water utilities and private water supply companies" (Upadhyay 2004), placing incredible pressure on already water-stressed rural areas and the women who cultivate them. Relatedly, water infrastructural projects have prioritized water delivery to factories, urban centers, and lucrative agricultural areas, sometimes causing communities to resort to makeshift infrastructure (Mehta 2020), such as in Vadabar, Gujarat, where villagers laid underground pipes to divert water from a nearby agricultural canal into an empty village pond for domestic use (Shah 2006). Deeming the water diversion illegal, canal authorities quickly filled the pipes with concrete, forcing local women to use a livestock water trough as a temporary water source.

The reality is, however, that women cultivators often make use of nearby canals and the dependability of their water flow for small crops, vegetable gardens, and orchard farming, even if using irrigation water for non-commercial agriculture is deemed illegal (Zwarteveen 1995). Rarely, however, is women's presence at the canal's edge noticed as signaling a citizen group's legitimate water need that could factor into the creation of formal irrigation water delivery schedules. Women's crops consequently slip under the radar of water management plans and designs, remaining invisible to decision-makers and excluded from irrigation statistics and sources that inform water policy and planning. Indeed, statistical indicators employed by regulatory reports to highlight regional water use measure visible, monetized water work and thus predominantly assess and forecast the water needs of industrial and large-scale agricultural sectors. The system trivializes water practices that do not visibly yield market value regardless of their vital importance to the continuation of local lives and livelihoods. Such gender-blindness suggests that water management is less the governance of resources within a geographic space as it is the governance of the female body as a spatializing effect. In short, water spaces become a construct of gender violence, with violence toward women taking place *as* space rather than *in* space.

Over the past ten years, feminist water research has pivoted from highlighting the absence of women in India's water management to suggesting ways of integrating women into water assessment and delivery mechanisms. Margreet Zwarteveen, for example, has issued suggestions and policy implications for the creation of gender-balanced irrigation infrastructure, including how to engineer, design, and operate systems with gender differences in mind; Lahiri-Dutt connects gender, water, and Indian law to translate women's everyday water management work into the formal legal economy; and Aditi Kapoor notes in research on gender-just climate adaptation that accurate statistical reflections of women's water work in national documents, such as India's National Water Policy and the 2021 census, are crucial to preparing agricultural and water sectors for climate impacts (Zwarteveen 1995, Lahiri-Dutt 2009, Kapoor 2011). To this scholarship, I add ecofeminist visualization as a way to assess, anticipate, and imagine the gendered organization of space through water use and water infrastructure. I suggest that gendering water management includes socioculturally re-envisioning the space of water through GIS maps. In what follows, I enact an ecofeminist visualization of "Water for All," a GIS map of Gujarat's plans for an ambitious hydroinfrastructural project to first contextualize the material and social consequences of gender-blind water management before envisioning GIS data as a condition of feminist action towards water equity.

"WATER FOR ALL"

In 2009, the Gujarat Infrastructure Development Board published a review of Blueprint for Infrastructure in Gujarat (BIG) 2020, a summary report of the state's integrated plans for the fast-tracked development of critical infrastructure across state sectors. Among the more ambitious projects outlined is the Kalpasar freshwater reservoir, the largest proposed dam project in the world, and an accompanying planned network of pipes and canals that would transport water to state districts from the reservoir's proposed location in the Gulf of Khambat. Under the overarching aim to ensure "water for all" (GIDB 2020) in an arid, water-scarce state, BIG 2020's vision for Gujarat's water sector included creating "systems and policies towards the effective, efficient and sustainable use of water to reduce poverty, improve human health and promote economic development" (GIDB 2020). Despite rhetorical gestures of inclusivity, no discussion is given to water equity or to the intense vulnerability of the rural poor in the face of increased competition for declining water supplies against growing sectoral demands and climate change. Instead, the construction and provision of water infrastructure is intimately tied to spatial articulations of state and economic power. For example, the water sector's main projects, visualized in figure 5.1, "Water for All," are heavily techno-managerial and investment-oriented, a fillip to Gujarat's economic goals to attain the Global North's infrastructure benchmarks. Against the backdrop of state districts, "Water for All" shows Gujarat's main water infrastructure, the Narmada Main Canal (thick, light blue lines), planned paths for new pipe networks (thin, dark blue lines), and the clustering of proposed water projects around Special Economic Zones (SEZs) (blue dots) and Special Investment Regions (SIRs) (blue ovals), the latter which describe proposed areas for immense, citylike hubs of economic activity. Crucial to the success of Gujarat's SEZs and SIRs are the ongoing plans for the Kalpasar reservoir, which sits, in a quasi-symbolic position, near the center of the map and is key to BIG 2020's economic goals. Measuring 30 kilometers (18.6 miles) and carrying 10,000 million cubic meters of fresh water, Kalpasar is slated to make available an additional 4,000 million liters of water per day for domestic and

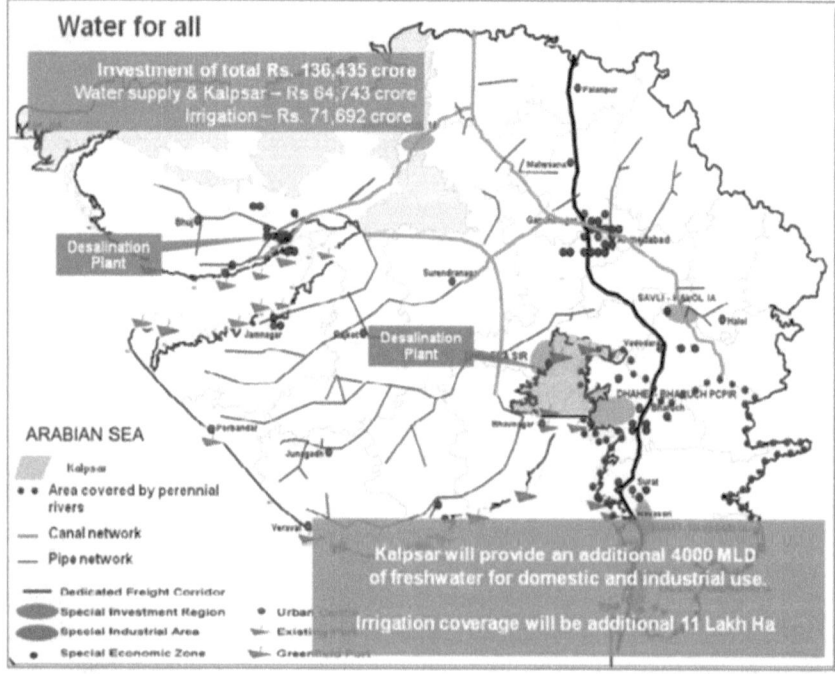

Figure 5.1. "Water for All."

industrial use and provide an extra 1,100,000 hectares of irrigation coverage to commercial agriculture through new canals and pipes.

Critical analyses of the dispossession and inequity resulting from hydroinfrastructures such as dams and reservoirs are well established, and while these critiques are essential to anticipating the uneven water development that might result from Kalpasar, in this section I want to locate the controversy of "Water for All" in a more holistic context. When taken together, the spatialization of women amid rural infrastructure and the scale of Kalpasar gesture toward state assumptions of the structural purity of water management practices. For one, the dismissal and statistical underenumeration of women's water work results from unspoken sociocultural mandates that all women's water experiences be striated within a domestic track. Evoking a similar linearity, "Water for All" envisions water management plans as one-directional relays that can turn on a "trickle-down" water source—one that that privileges rapid economic growth while supporting localities—and neatly reduce Gujarat's growing water crisis. The logic here, notably spanning both sociocultural and economic conceptual frameworks, discloses a disjunction between the anticipation and production of modern state space and its future, one that threatens water management's misunderstanding of itself as a system that functions apart from the social, ecological, and hydrological.

The infrastructural scale that "Water for All" visualizes and anticipates is shot through with the problem of incompatible human and nonhuman realities. In their introduction to *The Promise of Infrastructure* (2018), Hannah Appel, Nikhil Anand, and Akhil Gupta note that as "humans intervene in the climatic, geological, and evolutionary processes of the Anthropocene ... both the effects and future of modern infrastructuring projects appear increasingly tenuous" (Appel et al. 2018). Such tenuousness is symptomatic of an existing and worsening imbalance between nation-states' future development and the environmental thresholds of the landscapes that infrastructures depend upon and are situated within. Infrastructural megaprojects that marked the neoliberal race of national development and large-scale agriculture during the mid- to late 1900s have resulted in the need to engineer and closely surveille landscapes if these are to maintain their desired functionality. Ashley Carse, for example, demonstrates how watersheds in upland Panamanian forests need to be continually made and extended if they are to deliver reliable quantities of water to the Panama Canal (Carse 2014). Similarly, Stephen Lansing examines how the irrigation infrastructures of Bali's engineered rice terraces depend on artificial ecosystems to deliver water to nearly 20,000 hectares of paddy fields (Lansing 1991). In terms of Kalpasar, the existing tenuousness of built infrastructures has already placed an indefinite halt on reservoir plans, with the government of Gujarat announcing in 2018—two years shy of BIG 2020's decadal benchmark—that its feasibility is currently unclear

(Shah 2020). Not only is Kalpasar intended to monumentally span the east and west banks of the Gulf of Khambat, catching the freshwater of region's major rivers draining into the Arabian Sea, but the reservoir's main source of water, the Narmada River, has become increasingly polluted, its river bed and catchment destroyed by sand mining, its waters running dry in areas due to declining rains, droughts, and excessive damming (Counterview 2018, Aggarwal 2020).

The scale of India's water infrastructural tenuousness and development has caused climate scientists to call for a reconsideration of water management methodology. The engineering of landscapes to stockpile water as a resource is affecting the accuracy of regional climate models' hydrological projections. Accurate climate model projections have become crucial to formulating water management plans as stakeholders at the policy and water-management level strive to anticipate changes in India's water availability and needs. Reliable climate projection outputs, however, are impossible without water management taking a more nuanced approach to its own design and operation as the dramatic scales of water infrastructure—dams, canal networks, large-scale irrigation practices, and groundwater extraction—intervene on India's hydrology, compounding climate models' epistemic uncertainties and heavily skewing their representations of regional water futures. A 2014 study of climate impacts on India's summer monsoon, for example, reveal that a group of climate models misrepresented precipitation levels in the Tibetan Plateau and Hindu Kush–Karakoram–Himalayan range when their programming did not factor in the heavy, year-long agricultural irrigation activities that impact regional atmospheric circulation and contribute to the strength and spatial extent of monsoonal precipitations (Hasson et al. 2016, Levine 2013). Water management here becomes a physical variable in planetary hydrology. Under the regime of fast-paced economic growth, its practice and design impact not only the vulnerability of rural women but also the very materiality of atmospheric phenomenon.

Climate scientists have consequently sought to increase the accuracy of hydrological models by calling for stakeholders to fine-tune water management methodology. In a modeled assessment of climate impacts on India's upper Narmada River basin, climatologists Rickards et al. state that as river basins begin to face numerous climate-related managerial challenges and sectoral competition for water, "a more holistic methodology and long-term assessment are needed for water resources management across many Indian catchments" (Rickards et al. 2020, Mathur and AchutaRao 2020). Ideally, such a methodology would "incorporate anthropogenic basin interventions, such as water resource development projects, and account for population growth and demand from other water users, including industry and irrigated agriculture" (Rickards et al. 2020). To develop this well-rounded holism, the water sector would need to undergo intense self-scrutiny to assess its own specificities—design, operation, management, goals—which would include breaking down traditional social practices surrounding water use.

While Rickards et al. do not express a need for gender-just water governance, their call for a holistic methodology opens the very real opportunity for water management to recognize the gendered politics of water. Policymakers would, for example, be urged to closely investigate water uses across the formal and informal economy and would thus come in contact with the needs of female water users and the myriad ways in which women have always been "linked to the main canal" (Zwarteveen 1995), to use Zwarteveen's turn of phrase. Although there exist significant sociocultural hurdles to the realization of women as water actors, ecofeminist visualization is primed as a methodology to clarify the social formation of India's hydrology by spatializing women water users through GIS. Here, the political and not simply the ethical power of ecofeminist visualization becomes apparent and hails a gender-reckoning in water management.

READING DATA-BODY-SPACE

Socioculturally reenvisioning the space of water to usher in gendered water management includes practicing a unique spatial optics, one that brings the female body to the fore as an active force engaged with more-than-human bodies. Ecofeminist visualization provides a framework for this practice, as seen in the following visual analysis of figure 5.2, which presents a district-level mapping of female participation in agriculture compiled and imaged by Nitya Chanana-Nag and Pramod Aggarwal (2020) from India's last census. What results is an enriched mode of GIS reading that discloses data's affective capacity and fashions water users from within material relations, as opposed to legitimizing water users from the linear logic of public-domestic, formal-informal, man-woman binaries.

Figure 5.2 statistically visualizes the politics of difference that undergird India's water management practices. Each colored dot on the map marks 15,000 women engaged in agricultural production, with the red dots representing cultivators (women who own and tend to livestock and small-scale crops) and the blue dots representing laborers (women who work in the agriculture sector for income). In contrast to "Water for All," space in figure 5.2 is organized according to a *who* (i.e., women in agriculture) as opposed to a *what* (i.e., the production of water infrastructure for the subsequent production of national economic profit by male farmers). Using the map as testament to the role women play in India's agricultural production, Chanana-Nag and Aggarwal point to climate change hotspots on the map that are particularly

Figure 5.2. District level female participation in agriculture.

dense with women laborers and cultivators. As they do, they argue for the importance of equitable climate-smart agriculture to help reduce gendered climate risks. Though Chanana-Nag and Aggarwal do not discuss water use issues in their analysis, figure 5.2 indirectly maps women water users across India's 641 districts. While it is rare for women laborers to engage in irrigation activities due to the gendered dimensions of irrigation in formal agriculture, each cultivator is an informal water manager if she is to ensure a crop yield. The red dots therefore signal water uses and needs that remain unconsidered by water policies and water management authorities at local, district, and national scales, this is despite the fact that 60 to 70 percent of India's GDP results from subsistence agriculture planted, tended to, and irrigated by small-scale farmers, the majority of whom are women (Kapoor 2011).

Besides literally visualizing women's water work, figure 5.2 conceptualizes a layered interplay of data and material bodies. GIS maps are composed of a series of map layers (e.g., topography, hydrography, district boundaries, land use, water pipe networks), each of which represents a dataset that programmatically informs the construction of the final cartographic visual. While these layers together structure the GIS map, their dynamics are often considered a means to the visual end product despite the value of these layers as forces in themselves that, within the context of water management and gender equity, vibrate with political energy. Peeling off the top of the map, so to speak, to reveal the various layers and their interactions instigates a shift in GIS visualization. Instead of a planar distribution of elements in space, space itself becomes formalized as dimension and depth through the vertical layering of materialities. These layers provide an analytic with which to scrutinize the indices of materiality within the GIS map. The data points in figure 5.2, for example, transform into individual contact zones, with each point garnering conceptual weight not only from its status as a numerical marker but also from the density of material relations that it pinpoints, including the kinetic and generative tangling of bodies of women, water, communities, small-scale agricultural yields, and water infrastructure.

The density of the data point in figure 5.2 nudges against Stacy Alaimo's concept of transcorporeality, a theoretical space in which the human body becomes open to and crisscrossed by the more-than-human. Serving to critique, subvert, and evade "dominant modes of representation and ... gendered scenarios of visibility" (Alaimo 2010), transcorporeality describes a way of thinking across bodies or material sites that catalyzes, first, the recognition of the environment's ability to both affect and be affected, and, second, the realization that the human body is immanent within and inseparable from these dynamics. In her essay, "The Naked World," Alaimo exemplifies the power of this interchange in a reading of performances by environmental activists such as La Tigresa, a woman who strips to protest and stall logging operations in the old-growth forests of North America's Pacific Northwest. In her politics of undressing, La Tigresa discloses "a common corporeality, a shared vulnerability" (Alaimo 2010) between human and more-than-human bodies, issuing the ethical recognition that the human "is inescapably woven into a trans-corporeal, material realm" (Alaimo 2010). And while La Tigresa is only one of several examples of human bodies disclosing common, transcorporeal ground, the fact that hers is a woman's body presents transcorporeality as a method of re-engaging feminist analyses of the body with ecocriticism and environmental thought. In fact, in a statement that echoes Narayan's call for feminism's leave-taking of the discursive, Alaimo warns against scholars abandoning the fleshiness of materiality to engage with cultural and discursive reworkings of gender (Alaimo 2008). Like Narayan, Alaimo calls for feminist studies to pay closer attention to the specificities of bodies' embeddedness within material scenes and the ethical and political actions that can arise through their realization.

I want to deepen this specificity even further by adding data to the transcorporeal traffic of bodies. To practice ecofeminist visualization is to acknowledge that the data point within the GIS map does not plot bodies in space but alternatively formulates space through the transcorporeal contact zone that is its very existence. The blue and red dots in figure 5.2 are not simply visual articulations of data but concentrations of female bodies engaging with the bodies of the crops they care for and the bodies of water they use. This opens an epistemological "space" that acknowledges previously unseen material interchanges and newly formulates India's national space within the transcorporeal reality of women's agro-irrigation practices. Here, GIS data becomes an entity, and while not physically corporeal in the sense that Alaimo evokes, it riffs on the notion of bodies crossing and overlapping and itself takes on substance as a force that participates in this mixture. The data point affects GIS visualization by articulating a denser and more complex rendition of the socioenvironmental spaces that India's water management practices organize. In this new GIS optic, the point itself should not be considered a statistical takeaway but rather a conduit between the GIS viewer and the bodies mapped that conditions space through the viewer's response to the material interplay.

Notably, ecofeminist visualization is not a feminist theoretical exercise but, to borrow Yusoff's phrasing, "a realignment of sense through affective infrastructures, an affective mattering in the discourse of materiality and its worlds" (Yusoff 2018). As the feminist developmental economist Bina Agarwal writes, when providing a "Third World perspective on gender and the environment" (Agarwal 1992), women's and men's "relationship with nature needs to be understood as rooted in their material reality, in their specific forms of interaction with the environment" (Agarwal 1992). Ecofeminist visualization aims to visualize this rootedness in a recasting or expansion of what there is to see in a GIS map. It offers as praxis what I call an onto-methodology, a method of visualizing space so as to recondition the space of Being in anticipation of the future.

Such praxis will arguably become increasingly pertinent to feminist researchers and scholars interested in the gendered future of India's water politics due to the increasing feminization of agriculture. Over the past few years, male agricultural farmers across the country have begun to outmigrate to nearby urban areas, leaving their families behind to care for the farm and homestead as they search to supplement household income with work in the industrial sector. While Itishree Pattnaik et al. question whether this does not in fact reflect the "feminization of agrarian distress" (Pattnaik et al. 2018) due to the economic decline of the cultivator sector and degraded soil and water resources, the feminization of agriculture nevertheless suggests that the gendered composition of agricultural work is being reframed—and with it, the local governance of water. As women begin to assume decision-making authority and control of their own labor beyond the domestic threshold, women's irrigation activities and water needs will arguably become increasingly visible in the policy and public discourse surrounding water management. The onto-methodology of ecofeminist visualization can contribute to assessing the development of feminization in both agricultural and water sectors while empowering women by visualizing their work and emphasizing its importance to regional and national policymaking.

The political potential of ecofeminist visualization comes at a timely moment. India is currently strengthening its capacity for a national GIS infrastructure, with plans to develop a comprehensive GIS policy and instigate geo-data compatibility across all government sectors by 2025 (GOI 2011). The Ministry of Water Resources, for example, is expected to make available hydro-data that will guide developments in social wellbeing, urban and rural planning, and environmental health through GIS locating services, hydrological modelling, flood forecasting, and regional water needs and reservoir capacity estimations (GOI 2011). As both feminist theory and GIS method, ecofeminist visualization is primed to offer GIS and water management an embodied perspective of water practices crucial to the holistic development of India's future. In its arrangement of GIS optics, ecofeminist visualization settles the sense of sight into new formations charged with the material politics of the transcorporeal data point.

Though seemingly abstract, the imbricated materialities that data encapsulates—within both the data point and the literal layers of the GIS map—are not theoretical gestures as much as wayfinding markers that researchers and scholars might use to meaningfully incorporate women into water policies. Meeting the ongoing challenges of water scarcity means equitably and efficiently planning for communities' water needs. Ultimately, then, ecofeminist visualization is presented here as a political strategy, one that contributes to what Ariel Salleh calls the "pulse" of "women's efforts to save their livelihoods and strengthen their communities" through the entwined materialities of human and environmental systems (Salleh 1993).

REFERENCES

Agarwal, Bina. "The Gender and Environment Debate: Lessons from India." *Feminist Studies* 18, no. 1 (1992): 119–58.

Aggarwal, Mayank. "Narmada, the Forgotten River, Is Not On Anyone's Agenda." *Mongabay* (2020). Accessed November 25, 2020. https://india.mongabay.com/2019/04/narmada-the-forgotten-river-is-not-on-anyones-agenda/.

Alaimo, Stacy. "Trans-Corporeal Feminisms and the Ethical Space of Nature." In *Material Feminisms*, edited by Stacy Alaimo and Susan Hekman, 237–64. Bloomington: Indiana University Press, 2008.

———."The Naked World: The Trans-Corporeal Ethics of the Protesting Body." *Women & Performance: A Journal of Feminist Theory* 20, no. 1 (2010): 15–36.

Appel, Hannah, Nikhil Anand, and Akhil Gupta. "Introduction: Temporality, Politics, and the Promise of Infrastructure." In *The Politics of Infrastructure*, edited by Akhil Gupta Nikhil Anand, and Hannah Appel, 1–38. Durham: Duke University Press, 2018.

Barad, Karen. *Meeting the Universe Halfway: Quantum Physics and the Entanglement of Matter and Meaning*. Duke University Press, 2007.

Barry, Andrew. *Material Politics*. London: Wiley Blackwell, 2013.

Bosak, Keith, and Kathleen Schroeder. "Using Geographic Information Systems GIS for Gender and Development Development in Practice." *Development in Practice* 15, no. 2 (2005): 231–37.

Carse, Ashley. *Beyond the Big Ditch: Politics, Ecology, and Infrastructure at the Panama Canal*. Cambridge: MIT Press, 2014.

Chanana-Nag, Nitya, and Pramod K. Aggarwal. "Woman in Agriculture, and Climate Risks: Hotspots for Development." *Climate Change* 158, no. 1 (2020): 13–27.

Counterview. "Kalpasar Can't Be Implemented, Is a Non-Starter: Gujarat BJP's 'Tallest' Intellectual on Ambitious Rs 55,000 Project." *Counterview: Current Affairs* (2018). Accessed November 25, 2020. https://www.counterview.net/2018/05/kalpasar-cant-be-implemented-is-non.html.

ESRI-India. "Geo-Enabling Digital India." (2020). Accessed October 2, 2020. https://www.esri.in/esri-news/publication/vol8-issue3/articles/geo-enabling-digital-india.

GIDB. Final Report—Summary and Vision. Gujarat Infrastructure Development Board (GIDB) (2020). https://www.gidb.org/summary-big-2020.

GOI. "Establishment of 'National GIS' Under Indian National GIS Organization." Ministry of Earth Sciences, Government of India, 2011. doi.10.13140/2.1.3640.0644.

Haraway, Donna. *Simians, Cyborgs, and Women: The Reinvention of Nature*. New York: Routledge, 1991.

Hasson, Shabeh, Salvatore Pascale, Valerio Lucarini, and Jürgen Böhner. "Seasonal Cycle of Precipitation over Major River Basins in South and Southeast Asia: A Review of the CMP5 Climate Models Data for Present Climate and Future Climate Projections." Atmospheric Research 180, no. 1 (2016): 42–63. https://doi.org/10.1016/j.atmosres.2016.05.008.

Jiang, Hong. "Stories Remote Sensing Images Can Tell: Integrating Remote Sensing Analysis with Ethnographic Research in the Study of Cultural Landscapes." *Human Ecology* 31, no. 2 (2003): 213–32.

Joyce, Patrick, and Tony Bennett. "Introduction." In *Material Powers: Cultural Studies History and the Material Turn*, edited by Tony Bennett and Patrick Joyce, 1–21. New York: Routledge, 2010.

Kannabiran, Kalpana. "Feminist Deliberative Politics in India." In *Women's Movements in the Global Era: The Power of Local Feminisms*, edited by Amrita Basu, 119–56. Boulder, CO: Westview Press, 2010.

Kapoor, Aditi. *Engendering the Climate for Change: Policies and Practices for Gender-Just Adaptation*. New Delhi: Alternative Futures, 2011.

Keller, Evelyn. *Secrets of Life, Secrets of Death: Essays on Language, Gender, and Science*. New York: Routledge, 1992.

Kulkarni, Seema. "A Local Answer to a Global Mess: Women's Innovations to Secure Their Livelihoods." *Canadian Woman Studies* 21, no. 3 (2002): 196–202.

Kwan, Mei-Po. "Feminist Visualization: Re-Envisioning GIS as a Method in Feminist Geography." *Annals of the Association of American Geographers* 92, no. 4 (2002): 645–61.

Lahiri-Dutt, Kuntala. "Water, Women, and Rights." In *Water and the Law*, edited by Ramaswamy R. Iyer, 275–305. London, UK: SAGE Publications, 2009.

Lansing, Stephen J. *Priests and Programmers: Technologies of Power in the Engineered Landscape of Bali*. Princeton: Princeton University Press, 1991.

Levine, Richard C., Andrew G. Turner, Deepthi Marathayil, and Gill M. Martin. "The Role of Northern Arabian Sea Surface Temperature Biases in CMIP5 Model Simulations and Future Projections of Indian Summer Monsoon Rainfall." *Climate Dynamics* 41 (2013): 155–72.

Mathur, Roshni, and Krishna AchutaRao. "A Modelling Exploration of the Sensitivity of the India's Climate to Irrigation." *Climate Dynamics* 54 (2020): 1851–72. 10.1007/s00382-019-05090-8.

Mattingly, David, and Karen Falconer Al-Hindi. "Should Women Count? A Context for the Debate." *The Professional Geographer* 47, no. 4 (1995): 427–35.

McLafferty, Sara. "Women and GIS: Geospatial Technologies and Feminist Geographies." *Cartographica* 40, no. 4 (2005): 38–45.

Mehta, Avantika. "Drought-Hut Gujarat Has Water for Factories, but Not for Farmers." *IndiaSpend* (2020). Accessed November 23, 2020. https://www.indiaspend.com/drought-hit-gujarat-has-water-for-factories-but-not-for-farmers.

Narayan, Uma. "The Project of Feminist Epistemology: Perspectives from a Non-Western Feminist." In *The Feminist Standpoint Theory Reader: Intellectual and Political Controversies*, edited by Sandra Harding, 216–17. New York: Routledge, 2004.

Pattnaik, Itishree, Kuntala Lahiri-Dutt, Stewart Lockie, and Bill Pritchard. "The Feminization of Agriculture or the Feminization of Agrarian Distress? Tracking the Trajectory of Women in Agriculture in India." *Journal of the Asia Pacific Economy* 23, no. 1 (2017): 1–18.

Pavlovskaya, Marianna, and Kevin St. Martin. "Feminism and Geographic Information Systems: From a Missing Object to a Mapping Subject." *Geography Compass* 1, no. 3 (2007): 583–606. http://onlinelibrary.wiley.com/doi/10.1111/j.1749-8198.2007.00028.x/abstract.

Rickards, Nathan, Thomas Thomas, Alexandra Kaelin, Helen Houghton-Carr, Sharad K. Jain, Prabhash K. Mishra, Manish K. Nema, Harry Dixon, Mohammed M. Rahman, Robyn Horan, Alan Jenkins, and Gwyn Rees. "Understanding Future Water Challenges in a Highly Regulated Indian River Basin—Modelling the Impact of Climate Change on the Hydrology of the Upper Narmada." *Water* 12, no. 6 (2020): 1762. https://www.mdpi.com/2073-4441/12/6/1762.

Salleh, Ariel. "Forward." In *Ecocriticism*, by Maria Mies and Vandana Shiva. New York, NY: Zed Books, 1993.

Schultz, Nikolaj, and Bruno Latour. "Cosmology and Class: An Interview with Bruno Latour." *In the Moment, Critical Inquiry* (2020). Accessed October 2, 2020. https://critinq.wordpress.com/2020/01/13/cosmology-and-class-an-interview-with-bruno-latour-by-nikolaj-schultz/.

Shah, Anil. "Women and Water: Perceptions and Priorities in Rural India." In *Fluid Bonds: Views on Gender and Water*, edited by Kuntala Lahiri-Dutt, 172–84. Calcutta: STREE, 2006.

Shah, Jumana. "Gujarat Government Unclear on Kalpasar Dam Project After Spending Crores." *India Today* (2018). Accessed November 25, 2020. https://www.indiatoday.in/india/story/despite-spending-rs-30-71-cr-gujarat-govt-unclear-on-viability-of-kalpasar-dam-project-1174787-2018-02-21.

Upadhyay, Bhawana. *Gender Roles and Multiple Uses of Water in North Gujarat*. International Water Management Institute, 2004.

Yusoff, Kathryn. *A Billion Black Anthropocenes or None*. Minneapolis: University of Minnesota Press, 2018.

Zwarteveen, Margreet. "Linking Women to the Main Canal: Gender and Irrigation Management." International Institute for Environment and Development 54 (1995): 1–14. http://pubs.iied.org/6068IIED.html.

Chapter Six

Ecologies of the Digital Map
GIS and the Geography of Autopoietic Worlding

Erik Tate

"Environmental humanities" names an interdisciplinary matrix that is at once more specialized in its thematic focus than the traditional humanities, but broader and more interdisciplinary than ecocriticism or environmental history alone. Geographical information system (GIS) software, a powerful tool for storing, retrieving, visualizing, and analyzing all sorts of geospatial data, has enjoyed extensive application within the humanities as a methodological aide for studying history, literature, and culture. It is likely that a distinctly *environmental* humanities GIS (EH GIS) is emerging as an important methodological tool within a broader approach to the main fields of the environmental humanities: ecocriticism, environmental philosophy, and environmental history, not to mention subfields and study areas such as animal studies and environmental justice. As such, EH GIS could be considered as a distinct subtype of digital humanities GIS (DH GIS) and closely related, but more thematically focused than the practice of historical GIS (HGIS), which is widely used across geography departments and in historical research.

Several volumes have explored digital maps and GIS from cultural, historical, and critical perspectives, also drawing on critical social theory, poststructuralism, and posthumanist thinking (Bodenhamer et al. 2015), but few have addressed the possibilities GIS holds for the environmental humanities in particular. This problem has been compounded by the fact that despite increasing attention from humanities scholars to art and literature that engages with environmental disasters, growing ecological instability and the crucial importance of local responses to these threats, spatial and cartographic perspectives are only just beginning to be explored under the environmental humanities banner (Gladwin 2017). The environmental humanities, in other words, is still beholden to a temporal and historical frame. So, while the implications of spatial theory for geography and social justice scholarship have been explored, especially through the work of Michel Foucault, Henri Lefebvre, David Harvey, and Edward Soja, the role of mapping as a spatial practice and as a means of resistance to environmental injustice, remains underexplored. This chapter hopes to address this gap and to offer a theoretical lens through which to conceptualize the organizational, analytic, and rhetorical capacities of GIS with regard to environmentalist movements and grassroots sustainability projects. Two examples of GIS mapping projects will be discussed: Mark Terry's *Youth Climate Report* (Terry 2019), and Maya Lin's *What Is Missing?* (Lin 2019).

Academic work is building on two decades of participatory and collaborative uses of GIS mapping, for the purposes of spreading environmental awareness, promoting grassroots sustainability projects, and documenting the impact of and responses to climate change in local communities (Nicolosi, French, and Medina 2020). There is great potential for using GIS not only as a research tool but also as a means of producing openly accessible, online mapping projects that perform a range of aesthetic, ethical, and communicative functions in concert with deliberative and decision-making processes, sustainability initiatives, and environmental activism. Understanding the way that GIS interacts with social processes that shape space, and in turn shapes those same processes, will be of crucial importance for the future of the digital and environmental humanities.

The biological notion of "autopoiesis" can be traced to the Chilean cognitive biologists Humberto Maturana and Franciso Varela, who build on the conceptual resources developed in the realm of first-order cybernetics in order to help solve the problem of what constitutes a living system. Autopoietic systems are self-reproducing and self-organizing, exhibit organizational closure, and are structurally coupled with their environment and with other systems with which they share that environment (Maturana and Varela 1980). Autopoietic systems are living systems conceived as a network of relations of processes of self-production that are constantly renewing themselves and their various components, which in turn realize and sustain those very same processes over a given period of time. The prototypical example of such a system is the single eukaryotic cell, but autopoietic systems can also appear at much larger scales such as that of the human being or an entire ecosystem, depending on what sort of distinctions the observer initially makes and how the object is delimited. Living systems can be distinguished at multiple scales and are not defined by any particular material or structural property contained within a physical body, but rather by a particular type of organization composed of relations of processes of production. The network sustains the organism as a unity so that it is possible, for example, for all the cells in my body to be replaced with new ones without me losing my identity over time. Distinguishing between organization and structure is also important in this context. Organization refers to the network of relations of processes that define a system as a unified entity, whereas, structure denotes an actualization of an organization in a concrete, physical system composed of materials and properties, components, and spatial extension. Autopoietic systems are opposed to allopoietic, or other-producing, systems, for example, mechanical machines or an automobile factory (Maturana and Varela 1980).

This chapter proposes to integrate the concept of autopoiesis into an ecological framework for analyzing the participatory GIS project as a particular type of digital media object, a framework that has been applied in the context of other media such as film and interactive documentary. Sandra Gaudenzi has applied autopoietic (self-making) and cybernetic theory to interactive documentary by approaching it through the concept of the "living documentary," the notion that interactive documentaries are analogous with living biological systems in the sense that they spontaneously self-organize and are capable of continuous growth, renewal and self-maintenance (Gaudenzi 2013). Adrian Ivakhiv has formulated a process-relational approach to cinema in order to investigate the meaning-making and world-making capacities of film that unfold across three co-constitutive "ecologies": the mental, material, and social (Ivakhiv 2013). Together, these approaches enhance our understanding of films as objects or systems within a complex environment that weave together narrative, affective, and perceptual elements and shape important aspects of subjectivity and our being in the world.

GIS-based mapping projects may also be seen through the theoretical lenses of autopoietic systems theory and ecological world-making capacities. These perspectives are both founded in a process-relational ontology, which takes networks of differences and dynamic change across spaces and over time as the basic building blocks of objects and systems. As such, they are, like much recent critical cartography and GIS scholarship, committed to troubling the correspondence theory of truth underlying traditional cartography, a scientistic theory which holds that reality consists in knowledge of a pre-given, purely objective, independent world, and which draws a firm metaphysical line of separation between observers and their descriptions and representations of the world (the map) and the object of those representations (the territory). Leila Harris and Helen Hazen, for example, speak of fluid "map spaces" instead of fixed representations in order to emphasize a more-than-human perspective on "conservation-mapping practices as dynamic, performative interactions among people, landscapes, ecosystems, and species" (Harris and Hazen 2009, 50). If this is true for mapping in general, it must be all the more true of ongoing, participatory digital map projects that seek to inform, affect, and empower communities against ecological injustices. Both autopoiesis and the tri-ecological lenses can also be considered as postrepresentational approaches to knowledge creation that eschew many of the epistemological and ontological assumptions underlying traditional mapping practices. The next section considers this point in more detail.

CRITICAL GIS: POWER, PROCESS, SPACE

GIS is a relatively new technology. In the 1960s, at the behest of the Canadian government, Roger Tomlinson began to develop what would eventually become a tool for handling the massive amounts of data involved in inventorying and cataloguing the land and natural resources of Canada, resulting in the world's first computerized GIS (Zhu 2016). This concept and its technology was quickly and widely adopted and developed for a range of other uses across governments and private sectors. Insofar as GIS has been used by governments and corporations to catalogue and study land types, resources, animals, and ecosystems, it has largely been used in a managerial style to order, categorize, and domesticate space on the basis of a quantitative and empirical epistemology. In more critical terms, the dominant uses of GIS have often remained complicit in the continuing history of the importance of cartography for mapping new spaces for empire, state, and colonial expansion; sites for extracting resources; and in the service of capitalist development, as critical cartography has taken pains to point out at least since the foundational work of John Harley (2001).

As a result of the work of Harley and others, critical theories of cartography emerged in the 1990s that challenged the representationalist epistemology underlying traditional map production and use, critiquing the objectivist and positivist representational approach to mapping that has dominated cartographic science. What has been called the "epistemology of the grid" represents the dynamics of social space according to a rigid, static, compartmentalizing logic (Dixon and Jones 1998). It presupposes a Cartesian dualist ontology wedded to ocularcentrism, perspectivism that reinforces rigid distinctions between subject and object, production and consumption, representation and reality, cartographers and map users, and maps and social contexts (Casino and Hanna 2005).

Against this view, Harley and others investigated mapping as an ideologically laden discourse, text, or practice from a social constructivist perspective, which focused on the history and power relations behind the various beliefs and assumptions of classical cartography and the role of context in map-making practices (Dodge et al. 2009). Others emphasized the user and the context of use and consumption—how our active, bodily engagements with and knowledge of spaces also mediate our experiences of maps (Casino and Hanna 2005, 44). Seeking to augment and move beyond the social constructivism of critical cartography, some theorists began to rethink maps in terms of a postrepresentational frame and to move the focus beyond epistemological claims of maps to the ontology of mapping (Kitchin and Dodge 2007). Kitchin and Dodge argue that cartography should be seen primarily as a processual rather than representational science. They also question the ontological status of maps, proposing that maps be seen as incomplete, ongoing, and co-constructed through context-specific relations and practices (Kitchin and Dodge 2007). Maps are not fully determinate things but rather the site of unfolding relationships (Harris and Hazen 2009). Critical concern for GIS, also inspired by Harley's work, emerged alongside these developments during what has been called the "GIS wars," further emphasizing issues of positionality and reflexivity (Crampton 2010).

It is important to note that since the 1990s and 2000s, the power of GIS technology and its applications have become increasingly decentralized, whereas before it had been in the hands states, corporations, and trained technicians (Pavlovskaya 2018). The production of maps has come within reach of those traditionally considered merely as consumers of maps. There are those that have pointed to the potential for a more participatory GIS that emphasizes the co-production of knowledge between professionals and community members, giving them names such as public participatory GIS (PPGIS) (Mukherjee 2015) and participatory action mapping (PAM) (Bosse and Hankins 2017). Today, mapping technologies are less than ever the exclusive province of select experts with the requisite technical skills; now nonexperts can harness the power of GIS to produce accurate maps while community members themselves are better able to ask and pursue the key questions and set the research agenda. For example, Mary Terry and Maya Lin, documentary filmmaker and multimedia artist, respectively, have both created very distinct mapping projects addressing ecological issues that take advantage of this decentralization and democratization of access to GIS.

GIS scholarship has taken a process-relational turn in the wake of critical scholarship that has challenged the ontological assumptions behind the map and highlighted the contested histories of space and the power relations that shape the creation and use of geospatial technology and cartographic reasoning. Far from being neutral representation, GIS mapping continues to be entangled with power relations and epistemological assumptions, beginning with whether the underlying digital mapping platform that is chosen is provided open- and crowd- sourced, like Open Street Maps, or is proprietary, like Google My Maps. These have immediate consequences for the ownership of the data provided and the supporting technologies involved, compatibility with databases, who can access the data, and where it can be accessed. Production and design begins with such choices, but rarely ends there. Users also take part in the production process. That maps are not seen as finished products, but as ongoing, fluctuating, and open, is also a key insight of the autopoietic approach to digital media in general, and GIS-mapping projects in particular.

The author-user distinction becomes blurred in the context of autopoietic systems. Digital mapping projects afford increased levels of participation, meaning that users can upload content and voluntarily contribute data, thereby affecting the overall shape and content of the digital map and what others can see. This has a two-way effect on both the mapping project and the users themselves (Gaudenzi 2013). By collaborating with the GIS-mapping project on an ongoing basis, the user and mapping artefact reflects a converging and consistent reality. This ongoing, reciprocal feedback process of action and reaction, input and adjustment, is known in cybernetic terms as a feedback loop. The language of autopoiesis extends this arguments further; some actors may have privileged roles within the system, but at bottom, the living GIS map is characterized by spontaneous self-organization continually creating and enacting relations between a mass of heterogeneous parts: contributors and users, contexts and environments (both physical and digital), and social and technical systems.

From the perspective of critical GIS, this is known as a performative and ontogenetic understanding of maps and mapping (Dodge et al. 2009). Maps do work in and are worked on by the world. Mapping is recast as performative; the map is at the same time representation and a set of practices (Casino and Hanna 2005)—in fact, "the map" as a thing and "mapping" as an activity are no longer regarded as clearly distinct. The representational approach assumes that they are separate, such that the map is understood to be a stable product of mapping as a scientific practice. But this has serious epistemological drawbacks, not least of which is that fixed representation is static, whereas, the world is in a constant state of change, renewal and self-reproduction, growth and decay, formation and deformation, perturbation and compensation. The response is generally to strip away all of the "accidents," extraneous matter and movement, in order to represent the object as a stable entity, to capture the unchanging, idealized qualities of the object in question at the expense of its semiotic dynamism, contextual embeddedness, and material embodiment.

AN AUTOPOIETIC FRAMEWORK FOR GIS

The aforementioned autopoietic and ecological approaches are valuable as a bridge between the legacy of critical GIS and recent directions in the environmental humanities. On the one hand, collaborative GIS mapping projects can be thought of as autopoietic systems that have an internal capacity for self-organization and self-reproduction, as well as an open, interactive nature, which embeds it in the wider context of a map-using/producing community. Autopoiesis refers to the system's organization, the set of essential relationships that need to be produced and maintained in order that the system maintains its distinct identity, which is manifest in but also different from its actual structure and components. With autopoiesis, the interest is not in "properties of components, but in processes and relations between processes realized through components" (Maturana and Varela 1980). Its organization is the particular network of dynamic processes and relations that defines the type of system that it is and specifies its boundaries, while its structure and components designate the actual embodiment of these relations in a concrete—or in this case digital—space as this or that observable system. Mark Terry's *Youth Climate Report*, for example, consists of user-generated documentary film units stored on an online database and embedded as pins on a GIS map of the world (Terry 2020). The Geo-Doc, as it is called, is one of the most recent types of interactive, GIS-based documentaries to have emerged; it is a collaborative, multilinear, database and, in Gaudenzi's (2013) terms, a living documentary—it is alive in the sense that it is an autopoietic entity that grows and changes, yet maintains its organization against the structural changes growth incurs. The *Youth Climate Report* and other interactive mapping projects, such as Maya Lin's *What Is Missing?*, are fluid, processual, and relational and have a potentially unlimited capacity for dynamic growth, modularity in content and variability in overall structure (Terry 2019).

In order to establish this framework for understanding GIS-based, interactive mapping projects, it will be useful to gesture toward a classificatory scheme, specific to projects that are positioned at the intersection of social and environmental issues and that pursue or promote solutions to socioecological crises. These groupings represent different modes of autopoietic organization realized in GIS mapping projects; four basic families of interactive map according to their organizing function or goal can be discerned. These are: 1) The Expansion of Institutional Transparency and Public Accountability (for example, the *Land Matrix*, The University of Calgary's *Corporate Mapping Project*), 2) The Redistribution of Agency and Activist Empowerment (Mark Terry's *Youth Climate Report*), 3) Creative Remediation of Environmental Perception (Maya Lin's *What Is Missing?*, Leanne Alison and Jeremy Mendes's *Bear 71*), and 4) The Collection and Dissemination of Ecological Information and Scientific Knowledge (The IUCN's *Red List of Threatened Species*, IUCN *Red List of Ecosystems*, Global Footprint Network's "Ecological Footprint Explorer"). These groupings are based on which of the four types of autopoietic organization that each project manifests, but they are not meant as mutually exclusive categories. Communication is conceived here in terms of a digital ecology that comprises individual users; authors/curators; technical systems with inputs, outputs, and digital relays; and diverse, immaterial sign processes (semiosis). Instead of presenting an exhaustive list of types of projects, these groupings are a pragmatic way of classifying digital map-based environmentalist projects based on the kinds of affordances for thinking environmentally, perceiving ecologically, and collective action that they provide. This chapter will focus on categories two and three.

It is hoped that this framework will promote the further creation, exploration, and use of such online, environmentalist GIS mapping projects that address the complexity of climate change and ecological crisis from a more holistic theoretical vantage that is continuous with activist praxis. It is not only this chapter's intention to take "mapping" here in a narrow, literal sense, however, as in the actual creation of maps, but also to see maps as a broadly connected phenomena, embedded in human social, cultural and psychological processes, conditioning experience through cognitive, affective, semiotic, and preconscious, bodily dimensions.

The convergence of geospatial technologies and digital media forms progressively opens up a whole panorama of new possibilities and challenges for documenting and mapping, organizing and communicating knowledge, and effecting social and political change. Such convergences produce problems that critical and transdisciplinary knowledge-paradigms like critical geography and the environmental humanities are better equipped to solve. Many documentaries utilize maps and other tools for representing spatial relations to some degree. In some ways, the projects alluded to below take this to the extreme and show the continuity in form between the interactive documentary and the GIS mapping project. It is important to note that there are already many humanities GIS maps in existence, but in order to be relevant to the environmental humanities, historical GIS, digital humanities GIS, and other academically oriented digital humanities, GIS projects will largely be left aside (Levin 2011).

THINKING IN SYSTEMS: AUTOPOIESIS AND WORLD

The "Living Documentary" is a term coined by Sandra Gaudenzi in order to describe a theoretical lens for analyzing and understanding individual, rather than groupings of, interactive documentaries, and, as we will see, the main descriptor here, "living," also fits GIS-based mapping projects for reasons which will be made clear. Informed by Deleuze and Guattari's idea of assemblages and the second-order cybernetic biology of Maturana and Varela,

Gaudenzi defines a "Living Documentary" as "an assemblage composed by heterogeneous elements that are linked through modalities of interaction" (Gaudenzi 2013). The living documentary enacts multiple senses of the phrase "to be alive." An interactive documentary is alive in the same way that we might say a simple living system such as a single cell, a more complex system such as a single organism, or a superorganism such as an ant colony all constitute literal examples of living systems or assemblages thereof. But *live* can also connote something that is updated in real time, lively in the sense of moving and animated, or active in the sense of a live wire. Component systems can be isolated from the larger wholes in which they are embedded and regarded on their own, or they can be taken as a single entity; the choice is largely up to the observer, their goals or intentions, and the constitutive system-environment distinction that they initially make. While this lens might be satisfying in describing interactive documentaries that consist of a relational ecology of digital technologies, documentary filmmakers, users, and subject matter (Nash 2014), the integration of geographical and spatial reasoning as well as geospatial technology and GIS maps remains to be considered from an autopoietic lens. The Youth Climate Report Geo-Doc project, with its emphasis on environmental themes, multiple scales of representation, and global context, can ultimately contribute to a new spatial imaginary and cultivate social imaginaries of care, hope and resilience in the face of sociogenic ecological catastrophe and against the exploitative power configurations of neoliberal capitalism.

The ecological perspective on geographical information systems developed here is derived, in part, from the ideas of British cybernetician and anthropologist Gregory Bateson, ideas of critical importance for the environmental humanities. Traditionally, ecology has not fallen within the purview of humanities or social science scholarship, but rather of the hard sciences, as an approach to the study of relationships between organisms, and between organisms and their environment, quantified in terms of energy flows, forces, and physical effects. Despite being a great stride forward when compared with classical Darwinism, molecular biology, and population genetics, scientific ecology has still proven reductive in that it only recognizes material relations and leaves aside the questions of mind, meaning, communication, and the semiotic nature of information. The environmental humanities has emerged as broad domain of scholarship largely in response to scientific reductionism as well as to the historical partitioning of nature and culture into ontologically separate domains (Rose 1993). For Bateson, in order to have "an ecology" of mind, we must shift our focus from concrete, measurable effects to the more abstract, immaterial notion of "difference" (Bateson 1987). Unlike physical phenomena defined in the hard science way by relations of cause and effect, mental phenomena are caused by differences, which are transformed into information by, for example, an organism's nervous system. And the unit of information, according to Bateson, is the "difference that makes a difference" (Bateson 1987). Maps, likewise, do not represent territories but are composed of differences of difference. This brings us an important point about ecologies of mind and relations: insofar as maps are now primarily digital objects and produced by GIS, GIS can be seen as a *difference machine*, a perceptual apparatus that translates differences in the territory into differences on the digital maps.

The two approaches to GIS map projects considered here, the autopoietic and the ecological, are both significantly influenced by the systems thinking that emerged from cybernetics. The term itself, "cybernetics" (derived from the Greek *kybernetes*, meaning "steersman"), was coined in 1947 and popularized by Norbert Weiner in his book, *Cybernetics: Or Control and Communication in the Animal and the Machine* (Wiener 1961). Weiner, a mathematician by training, was working on an "anti-aircraft predictor," which could analyze the course of flight of aircrafts in order to predict probable changes in their position, and it would then use this information to improve the accuracy of an automated targeting system in order to shoot down planes or projectiles. The system, through the use of radar, would be able to scan the observed trajectory of the targeted aircraft, predict its course, and by repeating this process could continuously correct for errors as new information is fed into the system. This circular process is conceptualized in cybernetics as a negative feedback loop, which is defined as a circular control mechanism that systems use in order to correct for errors and to regulate the system as it pursues a clearly-defined goal, in this case to shoot down enemy aircrafts (Hayles 1999). Cyberneticists were interested in how goal-directed behavior in machines could be controlled and regulated by studying various kinds of feedback mechanisms, either in human-built machines or in nature, so that they could understand how machines could exhibit adaptive behaviors and improve their functioning by learning from trial and error.

A turning point came in cybernetic thought when Maturana, along with other key members of the Macy group of cyberneticists, published a paper concerning the visual perception of frogs in 1959 (Lettvin et al. 1959). The upshot of this paper was that the frog's perceptual system did not represent reality so much as actively construct it. In other words, the specific organization of the frog's cognitive and visual system determines what it sees. The assumption in cognitive research up until then had been one of straightforward scientific objectivity—that there was a cognitive situation in which an absolute reality "outside" and independent of the organism that somehow gets inside into cognition by way of a visual system. Maturana later reflected on the difficulties one gets into with the realist assumption that visual perception shows an independent reality and that this process can be observed "from the outside" by an observer. In a programmatic statement, Maturana writes that "everything said is said by an observer" (Maturana and Varela 1980), in order to bring focus to the fact that the observer itself, including the scientist, is not separate from the system being observed, but rather the epistemological situation resembles something more like observers observing observers.

Adopting a postrepresentational, organizational approach to knowledge guided by the concept of autopoiesis developed by Maturana and Varela, autopoeisis can be extended to virtual environments and highly dispersed systems mediated by digital technologies that connect a variety of users and locations. An autopoiesis extension of "digitization," with computers linked to globe-spanning information systems, massively expands the perceptual and action circuit of the human and has the potential to close the knowledge gap between science, activism, private organizations, and political decision-making apparatuses. Not without certain difficulties and contradictions, this progressive, emancipatory potential of digital and geographical information systems can be harnessed to produce positive social change and align those changes with human needs and the ecological services through which those needs are satisfied (Pavlovskaya 2018).

However, we must also understand that such information systems have hitherto been developed as a part of global capitalist systems and designed to serve the ongoing needs of empire, military intelligence, resource exploitation, and colonization. If the autopoietic lens is to have a critical edge, it must recognize postindustrial capitalism and its socioeconomic logic as the historical context in and for which these systems and technologies are created and deployed, and as that which often resists and contains any effort at meaningful social change. The language of autopoiesis is useful for describing units of the larger socioecological systems in which digitization occurs and, from the perspective of the larger system, of which it forms a component.

TRIADISM AND THE THREE ECOLOGIES

Digital maps can be analyzed and understood in terms of three overlapping ecological registers: the mental, the social, and the material. At stake from this perspective is the way maps weave together elements of each of the three ecologies and produce image-worlds, which in turn engender new modes of subjectivation and make possible new ways of being in the world, for both individuals and collectives.

The process-relational approach sketched above begins with the three overlapping ecological registers that Guattari distinguishes in *The Three Ecologies*: the environmental or material, the social, and the mental, or as Ivakhiv calls them, the geomorphic (inter-objective), the anthropomorphic (inter-subjective), and the biomorphic or animamorphic (inter-perceptual). Each of these ecologies contains implicit criteria by which GIS mapping projects can be evaluated (Ivakhiv 2013). Ivakhiv argues, after Heidegger, that cinema is "world-making" or "cosmomorphic" (Ivakhiv 2013). That is, cinema produces a version of each of the three ecologies in the form of three morphisms: geomorphism, anthropomorphism, and biomorphism. The world of the film has its own material, sociopolitical, and perceptual aspects, with the latter as the domain from which the other two are separated out and able to be regarded as distinct regions or domains of objects or subjects. This film world is encountered, of course, in the experience of viewing a film, which, likewise, has three aspects that Ivhakiv casts in the phenomenological categories of Charles Sanders Peirce: spectacle (firstness), narrativity (secondness), and signness (thirdness) (Ivakhiv 2013). Spectacle refers to the pure presence of the image of film, the unreflective immediacy of filmic experience; narrativity refers to the horizontal sequentiality of images and events, which is experienced as story; signness, or "exoreference," includes external references to the extra-filmic world or, for example, to real people or places, social or political events or circumstances, or intertextual references to other works, and this rounds off the meaning-making process in the filmic experience.

The third aspect of the process-relational analysis (after the mental and the social—the material) is the relationship between film and the world outside of film along with its materiality, its sociopolitical configurations, and its mental ecology. The first two are straightforward, but the third is less so. Material ecology extends over the material world, physical landscapes, ecosystems, and the traditional subject matter of scientific ecology. With reference to GIS, this ecology encompasses satellites, photographic equipment, radio waves and receivers, silicon microchips, electronic databases, circuits and information flows, computational devices, internet infrastructure, interfaces, screens, mobile devices and the whole assemblage of objects that constitutes geospatial technology. This domain comes to exist for each human being through perception and action in the world, bodily experience, and explorations, which results in our generating distinctions and oppositions between here and there, near and far, inside and outside, home and away, public and private. We learn about boundaries between spaces. Through the mobility of the body, we also learn what it is to cross and re-cross those boundaries, about journeys and homecomings. The geomorphic aspect of GIS mapping builds on this basis of experiences and knowledge of territories and territoriality.

The social ecology of GIS, on the other hand, includes the social relations among human beings who make and map the social world and circulate spatial knowledge. This domain includes politics, race, gender, class, and all the elements that perfuse social life and comprise differing social contexts in which spatial representations are put to use and take on different meanings.

GIS users and practitioners, as well as institutions like geography departments, corporations, governments, and other public and private interest groups comprise the social domain. The anthropomorphic productivity of GIS is not the way it extends human qualities to nonhumans, but concerns the politics of space, the places where different forms of sociality are permitted or prohibited, and the spaces of humans as distinct from the spaces of those considered as other, nonhuman, subhuman, inhuman, or monstrous.

Social and material ecologies are joined by a third register: mind or mental ecology. This is the central domain in which the subjective and objective elements arise and are encountered and from which the social and material ecologies emerge. The perceptual domain is the lived world in which subjects and objects, the social and the material, intermingle and interact, the experiential world of action and phenomenal appearances, affect, and meaning-making. Guattari models this after what anthropologist Gregory Bateson calls the "ecology of ideas," which "cannot be contained within the domain of the psychology of the individual, but organizes itself into systems or 'minds,' the boundaries of which no longer coincide with the participant individuals" (Guattari 2000). In this paraphrase of Bateson, Guattari interpolates the way ideas organize themselves into systems, suggesting that he had autopoiesis in mind. The three registers overlap; they are never encountered in pure form.

GIS mapping projects, as autopoietic unities, cut across all of these ecologies and draw upon each. GIS mapping projects incorporate elements of each register into an assemblage and unfold across each morphology. It takes up and transforms components of each ecological register into a unity of heterogeneous parts. This unity consists of an organization composed of relations between components, relations which are nevertheless independent of those components. Viewing the map project as an autopoietic assemblage helps to move beyond the somewhat circular movement between constructivism and criticism characterizing the cartographic tradition's treatment of the map as a text laden with ontological assumptions and invested with power/knowledge. An approach that is informed by social construction, Derridean deconstruction, and Foucauldian discourse analysis, such as that inspired by Brian Harley is helpful. Turning now to cybernetics and autopoietic media analysis, we will see that the GIS-mapping project produces autopoietic worlds in the same sense that film, for Ivakhiv, is "cosmomorphic" (Ivakhiv 2014), and as such produces worlds with their own version of the three ecologies.

AGENCY AND ACTIVIST EMPOWERMENT: *THE YOUTH CLIMATE REPORT* AND PARTICIPATORY GIS

The emergence of participatory mapping initiatives and the empowering of local communities to "speak truth to power" is at the forefront of recent developments in GIS enabled by the democratization of access to mapping technology and efforts to overcome the "digital divide" (Terry 2020, 136, 166). As Dodge and Perkins note, in a journal issue celebrating the twenty-fifth anniversary of the publication of Brian Harley's foundational essay, "Deconstructing the Map": "The move toward a more participatory approach to mapping, deploying GIS to empower, nowadays offers a way of integrating critique and construction into a praxis that was simply not available when Harley was alive" (Dodge and Perkins 2015). At the political helm of this development is *The Youth Climate Report* (YCR), which offers young and aspiring environmental activist filmmakers around the world the possibility of a direct line from their work on local environmental issues affecting their communities to the political movers and shakers of the United Nations and those who attend the yearly COP conferences on climate change. Geographical Documentaries (Geo-Doc) is a locative approach to interactive documentary filmmaking, and are "characterized primarily as a multilinear documentary film project presented on a platform of an interactive Geographic Information System map" (Terry 2020). Though the YCR is grounded in ideas and concepts emerging from a documentary tradition, its hybrid nature, I would argue, puts it on the border between interactive documentary and what is now being called digitally mediated participatory mapping (DGPM), of which the Geo-Doc could be considered an example.

DGPM is described as "a process of map-making that involves participation, making use of digital, often open-source, internet-based equipment" (Nicolosi et el. 2020), and is used by both researchers and activists to bring visibility and "reality" to both local grassroots sustainability initiatives and more distributed alternative economies. The YCR, as a Geo-Doc, also fulfills many of the functions of DGPM but emphasizes documentary film units as its expressive medium. This arguably endows the YCR with a more immediately affective dimension of aesthetic engagement that such DGPM projects as the Utah Resilience Map may lack. Its film units are short eco-docs that combine the indexical function of pins on a map with the iconic qualities of sounds and images, meaning that the YCR may also present visible evidence, rhetoric, narrative, argument and appeals to emotion. This likewise gives the YCR a relative autonomy and self-containment that change-makers may find useful in the course of deliberations.

Translated into the terms of autopoiesis, the YCR exhibits organizational closure. Organizational closure simply refers to the relative autonomy of an entity from its environment. While its organization is closed, it is structurally open, that is, it receives irritations or perturbations from its environment, but its organization, nevertheless, remains closed around a certain group of protected variables—in this case, its character and function as a climate change documentary, a tool for global youth to represent their communities around the world and to exert their influence on politicians, and a call for people to take action and create positive socioenvironmental change. Mapping, in the sense of DGPM, remains allopoietic (other producing) in the sense that its goal is to produce a map rather than reproduce itself, and the map it produces is predominately shaped by external factors rather than internal, organizational principles.

REMEDIATING ENVIRONMENTAL PERCEPTION: MAYA LIN'S *WHAT IS MISSING?*

Another project that follows this participatory path using mapping technology, but with emphasis on affective engagement with biodiversity and species loss, is Maya Lin's (2019) *What Is Missing?* project. This digital memorial takes the form of an interactive map covered with a color-coded array of clickable dots that contain various geo-located items, which include historical and contemporary accounts of local wildlife, environmental disasters and conservation efforts, images, videos and short user-submitted stories called "personal memories." It provides users with a chance to explore and engage with over five centuries of the "ecological history of the planet" (Lin 2012), but also encourages a retraining of the way complex environmental issues are perceived and the sorts of action that could contribute to mitigating extinction and biodiversity loss. The project emphasizes not only loss and despair, but also hope, and the "What you can do" section of the website recommends a range of straightforward lifestyle changes alongside information and statistics about the impact of current practices on the planet.

Maya Lin's project is both ecological, in Bateson and Guattari's sense of incorporating affective, social, and material dimensions of the human-environment relationship, and a living project. Each year, new content is added to the map from lay users, news articles, and other expert or academic sources. Likewise, the design qualifies as participatory because it allows users to contribute their personal memories, effectively adding emotional depth to the project by crowdsourcing personal stories about memory and loss. Maya Lin even hints toward the autopoietic organizational aspect of *What Is Missing?* by calling it "my last memorial, it is a memorial that will basically reinvent [itself]" (Lin 2012). As an immaterial monument to extinction, *What Is Missing?* exists in many places simultaneously and continually incorporates a range of components while remaining a distinct entity. As Gaudenzi puts it: "If autopoiesis puts the emphasis on logics of self-creation, on internal organization (what composes it) and on structural coupling (relations with the environment), assemblage puts the emphasis on heterogeneity of its components and co-relation with other assemblages" (Gaudenzi 2013). As an autopoietic assemblage of heterogenous components, *What Is Missing?* and other such environmental GIS mapping projects resemble living documentaries in their capacity to grow, change, and generate new modes of perceiving and interacting while maintaining a closed set of organizing principles that distinguish it from other systems and their components.

This constant reinvention and growth links to the ecological lenses adopted here in that this project does not separate environmental issues into discrete items as if they were to be solved one at a time, in isolation: "We tend to isolate and solve problems in a linear fashion, but the solutions that we need for the environment require a much more holistic approach" (Lin 2014). Likewise, Guattari attributes the cause of diminishing quality in social, mental, and "natural" relations to "a certain blindness to the erroneousness of dividing the Real into a number of discrete domains. It is quite wrong to make a distinction between action on the psyche, the socius and the environment" (Guattari 2000). The holistic or, for Guattari, "ecosophical" approach to the environment would regard GIS-based expressions as indissociably linked to the recognition of complexity, heterogeneity and a spirit of anti-reductionism that the YCR and *What Is Missing?* capture in diverse ways.

ENVIRONMENTAL HUMANITIES GIS AND THE DIGITAL ENVIRONMENTAL HUMANITIES

GIS maps have been used by academic historians and geographers to enhance their research since the 1980s (Gregory and Geddes 2014), and GIS software has also found a home among environmental historians. Though definitions of GIS abound, "it is broadly agreed that the key abilities of GIS are that it allows a geographical database to be created and the data in it to be manipulated, integrated, analysed and displayed" (Gregory and Ell 2007). Also important to note is the drift in historical GIS (HGIS) toward regarding GIS maps as a mere means for doing historical research, that is, a tendency to disappear the mediating influence of GIS technology away from critical scrutiny rather than to see GIS projects as objects of critical inquiry in and of themselves (to treat GIS as intermediary as opposed to a mediator, to use Bruno

Latour's terminology). Yet, despite appeals to "follow a GIS approach," there is no single method that defines GIS, and, as Jen Gieseking observes, "to describe GIS as a method implies that the software will collect the data and evidence for the research project and conduct the analysis" (Gieseking 2018).

Some are calling the area of convergence between digital technologies like GIS, digital map-making tools, and the environmental humanities by the compound label digital environmental humanities (EcoDH). As Stéfan Sinclair and Stephanie Posthumus observe: "Geographical Information Systems (GIS) are being integrated into the environmental humanities to create more complex mappings of cities, to track public awareness of environmental issues, and to critique dominant discourses about the nonhuman world" (Sinclair and Posthumus 2017). These mappings, consciousness-raising efforts, and critical evaluations are being used to expose and explore both the social, cultural, and economic bases of environmental problems and the complex interconnectedness of humans with nature. While sharing methodological and critical concerns for space, place, and digital technology, what distinguishes EH GIS mapping projects is the thematic foci adopted from the diverse perspectives comprising the environmental humanities, the methodological innovations, and critical concerns of the digital humanities, as well as the well-established approaches of the traditional humanities; thus, "the Digital Environmental Humanities potentially represent a sweet spot between conceptual and methodological concerns" (Sinclair and Posthumus 2017). The digital environmental humanities pushes forward the spatial, digital, and the affective "turns" in the humanities that have occurred over the past few decades, and could further benefit from drawing on critical cartography and critical GIS for valuable insights and examples to follow.

While there are large areas of overlap both conceptually and methodologically between these different systems and perspectives, it is likewise important to acknowledge the preconceptual and affective dimension of climate change and biocultural degradation: "An important project for the EcoDH, then, is to think more carefully about affect: the powerful, visceral, pre- or even non-cognitive feelings that arise and are transmitted in both virtual and actual environments" (Ladino 2018). What makes EH GIS unique is its potential to explore the complexity of such affectively charged entanglements between humans, technology, and multispecies environments, and the essential role affect (both positive and negative) plays in communicating pressing ecological issues to a broader audience of nonspecialists, encouraging social change and activist interventions into anti-ecological practices, and challenging anthropocentricism's twin pillars of socioecological degradation and metaphysical claims about human exceptionalism.

It is clear that the rapidly deteriorating diversity and complexity of these connections marking the Anthropocene era presents humanities scholars with an urgent challenge to work collaboratively with both the public and scholars from other disciplines, to address the impact of climate change and species extinction on communities from a social, cultural and critical perspective, and to give material and economic processes, nonhuman actors, and natural communication their full epistemological due—that is, interpreting and critiquing nature-as-text without reducing nature *to* text. It is this author's contention that EH GIS, autopoietic theory, and the three ecologies ecosophical approach can inspire new directions in environmental humanities research.

REFERENCES

Bateson, Gregory. *Steps to an Ecology of Mind: Collected Essays in Anthropology, Psychiatry, Evolution, and Epistemology*. Jason Aronson Inc., 1987.

Bodenhamer, David J., John Corrigan, and Trevor Harris, M. *The Spatial Humanities: GIS and the Future of Humanities Scholarship*. Indiana University Press, 2010.

———. *Deep Maps and Spatial Narratives*. Indiana University Press, 2015.

Bosse, Amber, and Katherine Hankins. "'These Maps Talk for Us': Participatory Action Mapping as Civic Engagement Practice." *The Professional Geographer* 70, nNo. 3 (2017): 1–8. https://doi.org/10.1080/00330124.2017.1366788.

Casino, Del J., and Stephen P. Hanna. "Beyond the 'Binaries': A Methodological Intervention for Interrogating Maps as Representational Practices." *ACME: An International Journal for Critical Geographies* 4, no. 1 (2005): 34–56.

Cosgrove, Denis E., and Carmen Cosgrove, P. *Apollo's Eye: A Cartographic Genealogy of the Earth in the Western Imagination*. JHU Press, 2001.

Crampton, Jeremy W. *The Political Mapping of Cyberspace*. Chicago: University of Chicago Press, 2003.

———. *Mapping: A Critical Introduction to Cartography and GIS*. Critical Introductions to Geography. Chichester: Wiley-Blackwell, 2010.

Deleuze, Gilles, and Félix Guattari. *A Thousand Plateaus*. Minnesota: University of Minnesota Press, 1987.

Dixon, D. P., and J. P. Jones. "My Dinner with Derrida, or Spatial Analysis and Poststructuralism Do Lunch." *Environment and Planning A: Economy and Space* 30, no. 2 (1998): 247–60. https://doi.org/10.1068/a300247.

Dodge, Martin, Rob Kitchin, and Chris R. Perkins. "Thinking About Maps." In *Rethinking Maps: New Frontiers in Cartographic Theory*, 1–25. London: Routledge, 2009.

Dodge, Martin, and Chris R. Perkins. "Reflecting on J.B. Harley's Influence and What He Missed in 'Deconstructing the Map.'" *Cartographica: The International Journal for Geographic Information and Geovisualization* 50, no. 1 (2015): 37–40.

Gaudenzi, Sandra. "The Living Documentary: From Representing Reality to Co-Creating Reality in Digital Interactive Documentary." PhD, Goldsmiths, University of London, 2013. http://research.gold.ac.uk/7997/.

Gieseking, Jen Jack. "Where Are We? The Method of Mapping with GIS in Digital Humanities." *American Quarterly* 70, no. 3 (2018): 641–48. https://doi.org/10.1353/aq.2018.0047.

Gladwin, Derek. *Ecological Exile: Spatial Injustice and Environmental Humanities.* Routledge, 2017.

Gregory, Ian N., and Alistair Geddes. *Toward Spatial Humanities: Historical GIS and Spatial History.* Indiana University Press, 2014.

Gregory, Ian N., and Paul S. Ell. *Historical GIS: Technologies, Methodologies, and Scholarship.* 1st edition. Cambridge; New York: Cambridge University Press, 2007.

Guattari, Félix. *The Three Ecologies.* Translated by Ian Pindar and Paul Sutton. London: Athlone Press, 2000.

Gutiérrez, Miren. "Crowdsourcing and Mapping Data for Humanitarianism." In *Data Activism and Social Change,* In Palgrave Studies in Communication for Social Change, 107–36. Springer, 2018.

Harley, J. B. "Deconstructing the Map." In *The New Nature of Maps: Essays in the History of Cartography*, 149–68. Baltimore: Johns Hopkins University Press, 2001.

Harris, Leila, and Helen Hazen. "Rethinking Maps from a More-Than-Human Perspective: Nature–Society, Mapping and Conservation Territories." In *Rethinking Maps: New Frontiers in Cartographic Theory*, In Routledge Studies in Human Geography, 50–67. Routledge, 2009.

Hayles, N. Katherine. *How We Became Posthuman: Virtual Bodies in Cybernetics, Literature, and Informatics.* ACLS Humanities e-Book. Chicago: University of Chicago Press, 1999.

Ivakhiv, Adrian J. *Ecologies of the Moving Image: Cinema, Affect, Nature.* Environmental Humanities Series. Ontario, Canada: Wilfrid Laurier University Press, 2013.

Kitchin, Rob, and Martin Dodge. "Rethinking Maps." *Progress in Human Geography* 31, no. 3 (2007): 331–44.

Knowles, Anne Kelly, and Amy Hillier. *Placing History: How Maps, Spatial Data, and GIS Are Changing Historical Scholarship.* ESRI Inc., 2008.

Ladino, Jennifer K. "What Is Missing? An Affective Digital Environmental Humanities." *Resilience: A Journal of the Environmental Humanities* 5, no. 2 (2018): 189–211.

Latour, Bruno. *We Have Never Been Modern.* Cambridge MA: Harvard University Press, 1993.

———. "Why Has Critique Run out of Steam? From Matters of Fact to Matters of Concern." *Critical Inquiry* 30, no. 2 (2004): 225–48. https://doi.org/10.1086/421123.

Lettvin, John, Humberto Maturana, Warren McCulloch, and Walter Pitts. "What the Frog's Eye Tells the Frog's Brain." Proceedings of the IRE 47, no. 11 (1959): 1940–51. https://doi.org/10.1109/JRPROC.1959.287207.

Levin, John. "DH GIS Projects." *Anterotesis* (blog). March 16, 2011. https://anterotesis.com/wordpress/mapping-resources/dh-gis-projects/.

Lin, Maya. "A Memorial to a Vanishing Natural World." Interview by Diane Toomey. Yale University, 2012. https://e360.yale.edu/features/maya_lin_a_memorial_to_a_vanishing_natural_world.

———. "Maya Lin Discusses *What Is Missing?* A Project and Interactive Website." Interview by Annie Godfrey Larmon, 2014. Accessed December 2, 2019. https://www.artforum.com/interviews/maya-lin-discusses-what-is-missing-a-project-and-interactive-website-49297.

———. *What Is Missing?* 2019. Accessed December 15, 2019. https://whatismissing.net/.

Luhmann, Niklas. *Social Systems*. Translated by John Bednarz and Dirk Baecker. Writing Science. Stanford, California: Stanford University Press, 1995.

Maturana, H. R., and F. J. Varela. *Autopoiesis and Cognition: The Realization of the Living*. Science & Business Media: Springer, 1980.

Mukherjee, Falguni. "Public Participatory GIS." *Geography Compass* 9, no. 7 (2015): 384–94. https://doi.org/10.1111/gec3.12223.

Nash, Kate. "Clicking on the World: Documentary Representation and Interactivity." In *New Documentary Ecologies: Emerging Platforms, Practices and Discourses*, edited by Kate Nash, Craig Hight, and Catherine Summerhayes, 50–66. London UK: Palgrave Macmillan, 2014.

Nicolosi, Emily, Jim French, and Richard Medina. "Add to the Map! Evaluating Digitally Mediated Participatory Mapping for Grassroots Sustainabilities." *The Geographical Journal* 186, no. 2 (2020): 142–55. https://doi.org/10.1111/geoj.12315.

Pavlovskaya, Marianna. "Critical GIS as a Tool for Social Transformation." *The Canadian Geographer/Le Géographe Canadien* 62, no. 1 (2018): 40–54. https://doi.org/10.1111/cag.12438.

Pickles, John. *Ground Truth: The Social Implications of Geographic Information Systems*. Mappings. New York: Guilford Press, 1995.

———. *A History of Spaces: Cartographic Reason, Mapping, and the Geo-Coded World*. Psychology Press, 2004.

Rose, Deborah Bird. "Thinking Through the Environment, Unsettling the Humanities." *Environmental Humanities* 1, no. 1 (1993): 1–5. https://doi.org/10.1215/22011919-3609940.

Sinclair, Stéfan, and Stephanie Posthumus. "Digital? Environmental: Humanities." *The Routledge Companion to the Environmental Humanities*, 369–77, 2017. https://doi.org/10.4324/9781315766355-49.

Terry, Mark. "Youth Climate Report." 2019. Accessed December 15, 2019. http://youthclimatereport.org/.

———. *The Geo-Doc: Geomedia, Documentary Film, and Social Change*. Cham, Switzerland: Springer, 2020.

Travis, Charles B. *Abstract Machine: Humanities GIS*. Redlands: ESRI Press, 2015.

Wiener, Norbert. *Cybernetics: Or Control and Communication in the Animal and the Machine*. 2nd ed. New York: MIT Press, 1961.

Wood, Denis, and John Fels. *The Power of Maps.* Guilford Press, 1992.
Zhu, Xuan. *GIS for Environmental Applications: A Practical Approach.* Routledge, 2016.

Chapter Seven

In the Retelling

Exploring Spatial Data as Narratives of Place

Michael Hewson

The question is this: do maps tell stories? Well, yes, they do, just not in so many words as a cursory glance at a map will admit. And like many storytelling gadgets—books, for example—maps have mostly migrated online. While digital maps in this internet age are now ubiquitous—some of us remember a time when maps needed extraordinary hand-eye coordination to fold them back into their storage shape, with the title page necessarily in the right place—has the internet age drawn books and maps closer together as storytelling devices?

The first digital mapping system that became coined *GIS* or geographic information system, is, as Terry expands on elsewhere in this volume, credited to Roger Tomlinson and the Canadian Land Inventory circa the mid-1960s. GIS turned paper-based cartographic symbols into computer-screen versions of the same thing—the digital media evolution of the map imposing new limitations, and creating new opportunities, for direction finding. Augmented by satellites, surveying and "story-maps," silos of several mapping disciplines have become an integrated geoinformation industry where GIS is the word-replacing, decision-making tool of choice (Konecny 2003).

The concept that a map can be a picture-enabled narrative has rattled around reading resources for some time. In the 1960s, and building on a centuries-old venerable map-making profession, Thomas Pickles (1960) hoped that his *cartographic interpretation* booklet would allow a map reader to turn a two-dimensional map into a three-dimensional mental picture of the landscape. Pickle's perspective is an interesting inversion, given the geodesic technical idea that a map results from projecting a three-dimensional Earth to a two-dimensional landscape visualization device—a flat Earth if you like.

Any map, including the GIS-enabled digital version, is a proxy for a landscape that can't immediately be seen. And maps have utility; they have, for example, the capacity to save us from being lost in the landscape, as evidenced by global positioning system (GPS) devices in the modern motor vehicle dashboard.

As digital mapping functionality evolves, GIS increasingly drapes the electronic Earth with a range of thematic data sets on which many strategic decisions are made: from Earth monitoring and ecological data; to social concerns such as health and emergency management; to property and infrastructure overlays; and a psychosis-inducing myriad of public service access (and the secret-squirrel world of political and military intelligence).

Pervasive now in many organizations, the GIS analyst interprets, manipulates, interpolates, extrapolates, or otherwise discerns layers of new meaning, new maps, new knowledge (perhaps even wisdom) on a landscape. Thus, a paper map has transitioned to a spatial information-based knowledge broker in the hands of GIS practitioners.

There is no end in sight for the usefulness of GIS where spatial data is ingested, arranged, analyzed, and formatted in ways that involve algorithms, prescriptions, and careful forethought, just like—well—writing! But beyond mere *writing*, is GIS successful with the function of narrative—the storytelling? Can GIS captivate, engage, suspend reality, and pass on the wisdom of elders to a reader?

SETTING THE SCENE

Maps have proven to be subversive over the centuries, and history shows that maps can usurp stories that society tells itself; just ask the genius of the mapping projection, Gerardus Mercator (1512–1594), who in 1544 was arrested for his mapping exploits on suspicion of heresy (Crane 2002). In the sixteenth century, the truth of his maps challenged some theological authoritatively held views on Earth's nature and shape—that perhaps the Earth was not the center of the universe. Mercator's maps had a truth-telling impact, just not the personal attention one would want if avoiding conflict with heavy-handed authority was a priority!

In defining the tribe—in providing community glue—stories of who we are and where we come from are essential. In recent times, GIS has been used to map oral traditions, to map foundation literature, and to map the modern-day audio-visual contribution to myth-making (Caquard and Cartwright 2014). Much has been written concerning the capacity of GIS to map literary landscapes (Alves and Queiroz 2015). From an emerging literary geography academic discipline, there is recent scholarship reinterpreting literature that maps the landscape in textual form—or perhaps better put, how literature can contain geographic facts (Ridanpaa 2018). Thomas (2013) explores the way maps inform literature by working together. Since scholars have already established that GIS can draw out fictional textual geographies (Taylor et al. 2018), this chapter takes that GIS-enabled literary geography utility for granted. However, very little literature discusses the idea that GIS output can *be* storytelling in and of itself. This chapter explores that concept.

Unless you skip the frontispiece as many might do, the first thing you read from the book *Ecocriticism* is a note from Garrard's editor (2012) who writes, "*Ecocriticism* explores the ways in which we imagine and portray the relationship between humans and the environment in all areas of cultural production, from Wordsworth and Thoreau through to Google Earth, J.M. Coetzee and Werner Herzog's *Grizzly Man*." The connections and intersections implied here are intriguing. So many geographical words! Firstly *exploring*, then, *the relationship between humans and the environment*. Ecocriticism is stamped here then as a work of geography. And so, the intersection of maps and literature has a foundation.

Reflect though that neogeography has become an information technology-enabled discipline of geographical enquiry. For example, remote sensing of environment images, whether by satellite, airplane, or remotely piloted aerial system (RPAS), extends early geographic information systems' collection of hardware components into many more mapping image vehicles and hosts. Consequently, the question becomes: how does digital mapping (such as Google Earth) do *exploring*? What narrative outcomes, what wisdom, accrues beyond technology? Once upon a time, exploration involved using a compass and chronometer, sextant and eye-glass, where wooden globes, parchment maps, and public lectures disseminated the garnered knowledge. The evolution of documenting exploration included the geometry enabled science of photogrammetry and aerial photographs and satellite-hosted global positioning systems. And consequently, civil infrastructure, epidemiology, agricultural crop health, and weapon delivery were better managed. But for the stories that sustain intergenerational society, does technology get in the way of maps telling Earth observation (for example) as a compelling narrative?

A ROAD MAP

This chapter compares the craft of GIS-based map-making with the art of creative writing. The development of cartographic practice is compared with the elements of imaginative writing. While the crafts of writing and mapping are variously described in the literature, this chapter leans heavily on contributions from Burroway (2015) for creative writing practice and Brewer (2005) to position the concepts of spatial storytelling.

For the sake of clarity, this chapter focuses on comparing the narrative creation of maps with the memes and norms of writing creative nonfiction (CNF). While many will argue that, mechanically speaking, writing CNF is no different from writing fiction, some of the knowledge synthesis in this chapter is more straightforward by selecting the CNF specialization. Essentially, the chapter assumes that while creative writing skills are similar for fiction and creative nonfiction, the latter is set aside by that essential, sometimes subtle, kernel of truth that readers expect in creative nonfiction (Perl and Schwartz 2014). Yes, maps can lie (as will be discussed), but a reasonableness test (what in Australia might be called *the pub test*) perceives maps as narrative nonfiction (creative or not). Indeed, the map-makers' inherent ethical practice around representing truth fits the norms and expectations of CNF like a glove. This chapter then compares the craft of cartography with the art of CNF imaginative writing and asks the map-maker to be a practitioner of narrative. Further, the chapter insists that a GIS user's attention to the craft of creative writing helps spatial and social science collaborate on compelling stories. Maps can indeed be storytelling devices.

If the research question is, *do maps tell stories?*, the next question becomes: how will any evidence accrue? Here the humanities hermeneutic process is deployed rather than a scientific method presenting observations and experiments. Hermeneutics posits that there are ways to communicate experiential truth that scientific methods cannot verify. The process of hermeneutics includes ways of interpreting documents so that personal meaning can be appropriated (Grayling 2019). This chapter reflects on the elements of both imaginative writing and map construction from a range of luminaries. The discussion is followed by four case studies, each observing how GIS map examples exhibit storytelling.

There is no difficulty describing digital maps as writing stories from a lexicographical perspective. The word geography originates from the Greek *geographia*, Latin for Earth (*gēo-*) and writing *(-graphia)*—thus *writing the world*. GIS is often described as the visual representation of spatial data. The *geography* part of GIS then tells stories about human interactions with the Earth, as El-Hadi (2020) suggests. Further, GIS is often described as the tool for making decisions using spatial data—this is the "information system" part of GIS. The core output of GIS, according to the combined acronym, is collating Earth system data to make a new Earth system narrative from the synthesized interactions and relationships between mapped parameter layers, superimposed as they are, one on the other.

Two, perhaps subliminal, central imperatives unfold (or are assumed) among this chapter's discussion comparing GIS to CNF. The first is that GIS can support Earth system meta-narratives such as Lovelock's (1979) Gaia-branded oversight of the Earth as a system. GIS is used to explore, document, and evaluate the Anthropocene's impact on the Earth system. The need to make peace with nature is perhaps the defining task of the current generation, according to a 2020 speech by Antonio Guterres, the United Nations Secretary General (UNEP 2021). GIS visualizes the evidence base for this imperative as its *core business*. Secondly, a simple acknowledgment is proffered by the author. The chapter recognizes that Indigenous Australians had (and have) the cultural capacity to map knowledge long before GIS or, indeed, paper maps. This spatial storytelling by song and artifact underpins an original terrestrial custodian's physiological responses to the landscape that colonizers and descendants cannot entirely appropriate or understand (Rigby et al. 2011).

The chapter examines the scaffolding constructs of CNF writing and cartographic practice and compares the two. The notion of the *narrative of place* is unpacked on the way through. The chapter then examines four vignettes, small reflective case studies, that muse over four mapped story arcs to illustrate how GIS can be CNF—how GIS can be storytelling—and thus, how GIS can be a tool for the disciplines of the humanities. The chapter builds on the discussion around narrative cartography headlined by Caquard and Cartwright (2014)—that maps *come to life*, not just in the creation process, but in the use of the map for a specific purpose, in a particular context—by personal abstraction.

CONSTRUCTS OF NARRATIVE CRAFT

There is much good advice on how to craft a narrative. Various mantras illustrate the advice: any story must *make one laugh, make one cry, surprise one*; a story must start with a *promise*, faithfully *progress* accordingly, and, finally, *pay off* for the reader's investment. More specifically, Burroway (2015) details key meta-elements of creative writing craft as *voice, character, setting,* and *story*. In many definitions of imaginative writing, the mandate to *show, don't tell* is abundantly clear. The mantra indicates that the text must deal with imagery, pictures created for the mind—by words. Thus, prose is a window to the world where the story helps the reader *see*. Further, a good tale allows the reader to suspend their belief—to release the power of imagination and accept the story's embellishments as acceptably accurate—for entertainment purposes. The idea of word *pictures* is thus already leaning to the possibility of maps to tell stories—avoiding the *picture is a thousand words* elephant in the room catch phrase for now.

The CNF craft of writing advice can be synthesized into several sections that serve to engage an audience. The first is the story structure or plot consistency (or story *arc*)—the writing imperative to fashion a story with some coherence and avoid ambiguity. Thus, a writer must make sure the scene transitions are smooth and don't interrupt the reader's reasoning powers. Secondly, a suitable narrative pace needs to be established, and the resulting rhythm needs constant time testing. Thirdly, the art of characterization and dialogue are tricky technical tasks that require much skill and editing perseverance to get right. And lastly, a collection of narrative digestion aids should not be neglected: choosing a point of view, crafting a narrative voice, a consistent well-chosen tone and pitch that collectively aids the storytelling. Thus, imaginative writing involves a set of rules, well before a writer picks up a pen or the digital equivalent. And here, the idea that an expert can (and does) break the rules to good effect is assumed.

Adding a layer of complexity is the notion that readers have subtle expectations of genre. Perl and Schwartz (2014) define the creative nonfiction genre as an umbrella that embraces several subgenres: memoir, personal essay, portrait, essay of place, narrative journalism, and opinion essay, or *op-eds* (opposite the editor) as they are known. Perl and Schwartz believe that creative nonfiction writing shares some key characteristics: an engaging personally authoritative voice; a narrative tension around self-discovery; remaining true to facts using expressive and effervescing imagery; interesting reader engagement tools (dialogue, description, metaphor, anecdote, character development); a finely honed freedom of rhythmic language; and a capability to let the narrative dissect the *big ideas*.

For instance, a CNF *op-ed* genre piece must set out a problem, rail with rhetorical opinion, and end with a solution that clinches the reader's conviction. Cameron Muir, with his tongue firmly planted in his cheek, observes that the craft of nature writing comes with some unwritten genre rules when he criticizes *writing nature by the numbers*, as mere recipes to "find a natural something, contemplate it, express awe, quote Thoreau, describe threats—end hopefully" (Schultz 2018, 214). Importantly when comparing GIS storytelling capacities, it is crucial to know that "CNF allows a reader space where reality is mediated and narrativised" (Perl and Schwartz 2014, 8). The abstraction inherent in mapping practice allows a map reader the same real estate.

To draw a "long-bow" abstraction for a moment, much has been written about the need for scientists to write narratively. Sword (2017) and Schultz (2009) are particular exponents of encouraging scientists to write less turgidly if the messages of an environmental crisis are to persuade public policy. Similarly, Olson (2015) distils the art of scientific storytelling to an acronym: ABT. Here ABT stands for *and, but, therefore*. For example, borrowing from fiction for the moment: AND a hobbit hordes a ring BUT is compelled to set off on a dangerous journey, THEREFORE saving the world as hobbits knew it. If GIS can do ABT, then perhaps the case is closed concerning the capacity of GIS to be a storytelling tool. Back to that shortly!

Narratives of Place

The CNF subgenre, the *essay of place,* nominally provides the most direct link between the written and the mapped narrative expression. The *essay of place* is where the landscape or the scene, perhaps a series of scenes, is the character or the plot. Writers of this CNF subgenre combine the facts of the place with vivid sensory descriptions. A map, GIS derived or not, is a visual essay of a place. In a map, the place is the character.

CONSTRUCTS OF MAPPING CRAFT

For making maps, the equivalent of the craft of imaginative writing is in a professional practice known as *cartography*, or in conventions known as *cartographic elements*, not that all GIS skilled map-makers would recognize the term; perhaps *graphic design* is a better-known synonym. For some, cartographic rules are most clearly defined for producing topographical maps, where paper map conventions were incorporated into the GIS computer-generated imagery design concepts as intentional narrative abstraction devices. Cartography is defined more broadly as the framework for map-making, the principles, techniques, and standards for optimizing maps as spatial communication devices (Mayhew 1997). Like creative writing, there are no single international (or, indeed, national) standards for cartography per se. There are international standards for the data store mechanisms and protocols behind modern GIS, but that is becoming too technical for this present chapter. More practically, many companies stipulate mapping practices and standards to aid consistent story prosecution. These standards vary but often contain similar-looking symbology. And so, like creative writing, there is a certain established *word-of-mouth* set of norms and memes for making maps resplendent with cartographic elements.

Australia's Intergovernmental Committee on Surveying and Mapping (ICSM) lists one set of cartographic considerations and map-making specifications (ICSM 2020). The list includes the size of the map and the importance of the scale to be used (scale: what map standard length represents on the actual Earth); the layout of the map elements on a page and the *marginalia* that is to be used (the mapping items that inhabit the map edges, or *margins*, to be described shortly); the use of text and symbols, colors and patterns (e.g., how terrain height is visualized); and the use of contextual map insets.

Arguably the most detailed map presentation norms were consolidated with the acronym DOGSTAILS (as familiar to some map-makers as the yellow rectangle trademark of *National Geographic*—for others, the organization from which DOGSTAILS emerged in a popular form). Rearranged in order of importance, TODALSIGS means: the margins of a map need a title (what, where and when); orientation (perhaps a compass rose); a date when the map was made; an author who made the map (to assess authority—or someone to blame); a legend to explain the symbols; scale; an index where specific information can be located; and perhaps a grid that aids in the context of location and sources where map information was provided. Taken together, these conventions aid storytelling.

Some map-makers have several other critical imperatives on the GIS map design craft checklist. Brewer (2005), for example, lists the quite technical mapping projections and layout hierarchy as important map design craft. Map design consists of careful consideration for balancing the map elements; space, label and symbology, placement, size, extent, and color decisions (of all sorts). GIS map design goes far beyond topographic map design these days, and consideration for a raft of new maps, tables, charts, and spatial information presentation norms are well documented.

In a GIS, a *place* can have color, texture, smell, and a defined character through a range of cartographic conventions and related attribute data. Some of that attribute data can become symbols and labels signposting the visualization of character. The cartographic attention to symbology and labeling color hue, saturation, and intensity provide the sensory detail that is the hallmark of the literary essay of place. Indeed, now that GIS can display three-dimensional images, place visualization capabilities are richly enhanced.

Whether cartography elements (however defined) are storytelling devices will be discussed by looking at the four following case studies of contemporary mapping experience.

GIS AS A NARRATIVE

Perl and Schwartz (2014) define the *voice* of creative nonfiction as a friend-to-friend relationship—a dialogue where the writer sounds genuine and possesses integrity. The story conveys personality, emotion, and color. A set of facts in a report devoid of a narrative is not a story—but a report. For example, a mere description of virus transmission trends for a given nation is not the story. However, the narrative journalism piece weaving a sensory tale of the plight of nursing home residents and their attending caregivers as a day unfolds is a story with voice, characters, protagonists, story arc, and engaging narrative.

What then is the GIS's *voice*? Is a GIS map, perhaps, just a report? Like the report/story dichotomy, the answer depends on the map construction. Maybe the answer depends on how one reads a map. A topographical map unfolds for the bushwalker, and a rail-line map extends meaning for the train-spotter. The user engagement level depends on the map user's motivation to put the map to a purpose. A map showing the statewide count of cholera cases per mesh block or government boundary is a report. A map showing the statewide count of cholera cases with symbols or colors highlighting the graduations of disease concentration against a measure of wealth and relative privilege is a story. Anthamatten (2021) discusses qualitative visual variables (shape, orientation, color hue) and quantitative visual variables (size, texture, color value, and saturation) as means of telling stories via thematic maps. Thematic maps are a conversation between the map-maker and the map user because the cartographic elements are visual cues to the meaning of the map and its message.

GIS practitioners take a set of GIS mapping elements—a stack of disparate data with a unifying location if you like—and craft a story from arranging that stack according to how the map-maker envisages a story arc. A GIS practitioner starts the pile at the bottom with a "base map." A base map is defined as the background reference layer onto which subsequent map layers are placed (Law and Collins 2020). The base map is often a canned topographical or thematic map, a satellite image or aerial photograph, or a derived product—such as a digital elevation model to give the topography shape and shading. This spatial reference layer provides the narrative setting: the context on which the mapping chronicle unfolds. Cartographers know that consideration for the base map's visibility, color, and impact sets the foundation for all the plot turns that follow in the pancaking of the story arc, the layers of cartographic detail.

The ethics of creative nonfiction writing craft pins authors to spin a yarn around a kernel of truth. The extent to which the truth is subsequently bent, shaped, painted, or clothed depends on where the writer places their work on the creative to speculative nonfiction spectrum. For GIS narratives, the base map is that kernel of truth. The map-maker can explore the creativity and subjectivity of presenting the story with the placement, manipulation, and presentation of the future mapping layers.

And yet, maps can lie.

Ab initio students of GIS skills acquisition are drilled to avoid map fables associated with presenting certain thematic information. The point is made most classically by comparing the thematic map of the total population for a city, where the choropleth (color-shading scale) provides the population's visual narrative per census block as a population number graduation from light to dark. More minor population counts are lighter in intensity. The fable appears due to the spatial fact that census blocks are of varying size, so a large block with a small population can have the same classification color as a much smaller block with the same population size. Typically, this makes the population density map (which it isn't yet—but bear with me) look deceptively heterogenous. There appears to be a misleading pattern to the population density. As foreshadowed, the map of population counts per census block is not a *density*. Yet *density* is what a map reader thinks they want to see when presented with a population map. There is something in a neurological makeup that dictates this habit. Students are tutored to solve the map reader disconnect problem by creating a population map as a ratio: the population divided by area. Thus, a population density map now contains the knowledge expected: a clustering of inner-city census block greater population densities than those for the peri-urban census blocks. The map-makers' truth-telling needs to consider normalization to aid the reader's perception of the narrative.

The plot, the story-arc, the saga that unfolds through the map's interpretation is as carefully planned in GIS by the cartographer as it would by a narrative journalist (or any other CNF category writer). As Brewer (2005) notes, attention to GIS map design and building a polished layout will help the map reader understand its message.

And if maps can lie, then they can undoubtedly distort reality. But this distortion, this abstraction, can be put to narrative use. A mapping enhancement (a positive spin on distortion) can be analogous to the literary imperative to *show, don't tell*. Engaging a reader is a risky business—the capacity to keep someone glued to the pages, wide-eyed and heart-thumping, is hard-won. Many books are thumped on the bedside table when readers lose the will to read. Technical cartographers crave attention to geographic reference frameworks, projections, scale, and the underlying geometry, while perhaps neglecting the need to create engagement. GIS readers intuitively understand that map distortion, such as a misrepresentation of scale, can (if cleverly crafted) enhance the narrative and provide impact to enliven the message, just like creative writing. It takes some soul searching for a map maker to loosen up and embrace embellished scales. But once in that zone, the mapping story can leap to another level of impacting visualization. To return to the mapping of population size, a two-dimensional projection of the Earth, where each country is sized and colored according to its population size, presents a far different map than where the country boundary location is accurately established. Examples of such narrative visualizations, the distortion of GIS maps for narrative impact, are provided by Reuschel et al. (2014).

As Brewer (2005) discusses at length, map makers select color schemes, color space (whether CYMK, HSI, or RGB), the width of lines, internal polygon hatching, or a myriad of shape representations to tell a story by map. The expansive design choices for labels and symbols available to GIS users are enough to invoke professional psychology problems—a GIS practitioner has so much choice as to experience a real decision dilemma! And here DOGSTAILS start to wag—the practice where graphical and textual elements of map-making converge in the endeavor of storytelling. The balance of marginalia and cartographic elements (title, north arrow, scale bars, legends, explanatory texts and grid representations) play their parts as narrative elements. Here, in the map, the graphical elements do the heavy narrative lifting. For example, too much white space on the map and the reader perceives a story problem—what's missing? What's out of kilter?

The careful selection of GIS parameter (or database "field") symbology (such as the ramp of color hues to visualize parameter strength) allows mapping emotions. Maps can quantify the range of dimensionless aesthetic qualities such as "beautiful" or "sublime" with color shade or intensity change, leading to new insights and interpretations (Taylor et al. 2018). Clearly, visual reading is different from textual reading. The question is then: is one better than the other? Arguably, for the purposes of narrative, just *different*, perhaps *complementary*.

Pearce (2008) argues that a cartographer has to work at their craft to convey a sense of place in, for example, cultural or historical mapping. GIS-based storytellers need to be attentive to the narrative functions of the technology. Sui (2015) argues that the technical GIS focus on cartography, quantitative data collection, computer, and databases should be alleviated by user attention to six senses of the mind (design, story, symphony, empathy, play, and meaning) if GIS is to be engagingly narrative.

An associated GIS practitioner skill of making new knowledge by combining GIS data layers is known as *spatial analysis*. Longley et al. (2015) determine that spatial analysis tools are the heart of the GIS. The collection of spatial analysis "picks and shovels" applies a range of transformations and manipulations to add value, reveal patterns, and support decision-making from what might seem on the surface like disparate spatial data.

The skilled spatial analyst can make simple, single GIS layers into new knowledge that builds a narrative tension that, in the right hands, builds suspense and creates the essential narrative climax (for a map to truly be a story). GIS does maps AND maps can meld spatial entities (layers) to a cogent procession, BUT GIS spatial analysis tools can create new knowledge from disparate sources AND solve intricate spatial data problems, THEREFORE integrate mapping layers into a narrative. Bingo—an earlier ABT story arc realized!

VIGNETTE #1—LIES, DAMNED LIES, AND MAPS

The fiction part of the term *creative nonfiction* is the license to use storytelling tools to enliven, perhaps embellish, truth-telling. So, the story bends to accommodate the imagination and thus stretches to engage the reader. A key question for the CNF author is this: when does the truth get so twisted that it breaks into a lie like a pounding wave of surf thumping the beach? This question stretches the ethics for narrative journalists and other keepers of strategic stories alike. And like statistics, maps can lie. Sometimes the lie, or at least the misdirection, can be entirely deliberate for the sake of storytelling. On the other hand, mapping misconceptions can be imparted by lazy map creation—or lack of attention to good narrative practice.

The Australian summer of 2019/2020 will be long remembered for the most intensive and extensive bushfire outbreaks in living memory. Some 4.3 million hectares of Australian land cover was burned. Around 37 percent of the national park estate for the Australian state of New South Wales was burned. An eye-watering 57 percent of the Australian population was, to varying extents, exposed to bushfire smoke (Hughes 2021). Australia's rural firefighting organization chiefs met to review the fire season as the summer passed. As a group of elders, they remarked that the fire season had been unprecedented to news media outlets. The word "unprecedented" was to become the catch-cry of 2020, as it occupied the Australian emergency public briefing material for many long hot months before the term became ubiquitous as the COVID-19 exploded around the globe.

And GIS played a significant role in providing spatial information for fire operations, but what about the story for engagement with the population? How did that go?

The Western Australian State Government's land information authority published a bushfire site map known as "MyFireWatch" to provide a public information service. On a map of Australia, each bushfire "hotspot," determined by a satellite image–processing algorithm, was marked by a single fire flame representation. A symbol on the map designed to tell the development story of the unfolding catastrophe.

When zoomed to national scale via a web-hosted GIS system, the map showed a whole swag of flame icons that came across to a viewer as if most of the Australian landmass was on fire. When you consider that Australia is around the size of the continental United States of America, such a map seemed seriously scary. When a user changed the map's scale from national to regional then local scale, the flame icons indicated a more reasonable fire-ravaged extent. The map's problem is that the fire icon stayed the same size regardless of the viewer's zoom level. This, of course, gave most of the public map users an eye-popping impression that the entire continent was one immense conflagration.

The Landgate website terms and conditions clarify that the fire hotspot mapping had many caveats, such as the icon's color does not represent the fire's severity. A skilled user of the Landgate MyFireWatch site can retrieve beneficial bushfire information, including facilities for comparing vegetation greenness index, burned areas for previous years, and lightning activity. But a *public-good* map needs to provide information to the casual, untrained website user. First glances fill a mind's perception that later critical thinking can change. The narrative has become a neuron-enabled memory. You get one chance on the internet sometimes. The storyteller's symbology selection needs careful thought for public information narrative purposes.

A hallmark of trusted creative nonfiction is a reader's trust that some kernel of truth exists. With a map, that truth is graphically portrayed in the presenting symbology.

VIGNETTE #2—TREND VOLATILITY MAPS

Maps are used in the visual media to accompany news stories with a spatial context, and often, these news stories are severe weather-related events. In recent years, maps have narrativized government elections because of the geographical aspects of the election of political figures. As the year 2020 came to a close, many online news organizations (the Australian Broadcasting Commission included) tracked the presidential election of the United States of America with various map forms. In all cases, the US state boundaries were abstracted to symbolize the voting trends—multiple ways of showing the results for the Democrat (blue) and Republican (red) electoral college vote numbers per state.

The most compelling of these maps was a version that indicated the size of the electoral college votes as several hexagons per state. The larger states had more hexagons. While each state boundary was necessarily distorted, the overall creative arrangement made for a recognizable map of the continental United States. The maps were a creative visualization of the college vote trends as red and blue hexagons ebbed and flowed across the news site graphic. The narrative was compelling viewing, and (for the present author at least) the presentation stood out as an innovative mapping-enabled narrative.

The map abstraction was a proxy for the truth of the election trend and valuable to the storytelling. The various symbol elements made the story come alive—the use of color, the hexagon abstraction, and yet the *big picture* could be readily perceived at a single glance.

The web-hosted maps provided additional facilities to drill down into the data so that a user could find the percentage votes the data changed, electoral booth by booth. Thus, the presented graphics provided the mechanism to select more narrative details.

It needs to be said that a reasonable CNF demands, indeed requires, a satisfactory ending. Sadly, maps as output cannot guarantee such a satisfactory outcome for everyone!

VIGNETTE #3—MAPS OF MASSACRE

Maps can provide an illustrative enhancement to confront society's ills and disrupt the story society tells itself. In 2017 using ESRI ArcGIS web hosting services, the University of Newcastle (New South Wales, Australia) hosted a so-called *frontier wars* map. These wars were the massacres of Australian Indigenous Peoples, and consequent tit-for-tat reprisals, during the European colonization of unceded Australian lands from 1788. The map (Ryan et al. 2017) is a collaboration between humanities scholars and the Australian Institute of Aboriginal and Torres Strait Islander Studies being updated with additional information through to 2020.

The map is a simple design—a dot for an approximate location of a historically verified massacre site (acknowledging the complex causal cultural contexts) on a base map of Australia. A map user can select a location and drill into the research team's narrative details.

The maps make for confronting narrative. The number of dots, the widespread location, the sheer weight of yellow dots (European initiation/reprisals) versus blue dots (Indigenous Peoples reprisals) is visual narrative devices that need no textual embellishment. The compelling story super power of this map is the visualization of human truth. The spatial extent of simple dots is the story arc climax that hits you hard—right between the eyes.

The simple user interaction of the map is intriguing. A user can choose a date range to query the map's stark narrative function—triaging for *when* among the *where*. The map allows users to pick a period and advance that period through the site database, creating a time-series movie of massacre events. The visualization of dots appearing over time shows the expansion of the colonizers in a far more engaging way than reading a table of historical dates and facts.

The researchers note that the map and its underlying historical attributes are the first Australia-wide frontier war record. The map timeline indicates that massacres spread steadily across the Australian colonial frontier with notable peaks from the 1820s to 1840s in the southeast of the continent, the 1860s and 1870s in Queensland, and the 1890s in the Northern Territory and the Kimberley region in Western Australia, marking the march of colonial carnage.

While the generally accepted birthplace of GIS was within a European cultural heritage circa the 1960s, Indigenous Australians, with a continuous lineage of country stewardship in excess of 60,000 years, have a history of representing the spatial aspects of respecting and utilizing sustainable country in their cultural practice, particularly artwork, often with the replication of symbols, just like modern cartographers. In a foreword to the *Macquarie Atlas of Indigenous Australia* (Arthur and Morphy 2019), Patrick Dodson, Indigenous Australian and senator for Western Australia writes, "maps are important to me. I love the way they tell stories about Country. I read maps closely, looking at the way the map maker has chosen to draw the past and the present; the ways in which they have given weight to various elements of the physical, the cultural and the historical; and the ways in which they tell the stories of First Nations people."

In a culture that hands down knowledge, education, culture, meaning, advice, direction, religion, social cohesion, medicine, sustenance, sustainability, and right relationships in song and story, and artifact, maps are in the mind. Maps are the land imagined and memorized to remember meaning for constant use (Neale and Kelly 2020). The term "song-lines" has been used to encapsulate the concept for Indigenous Australians, but even that term is a non-Indigenous summary grappling with a complex idea.

VIGNETTE #4—VEGETATION HEALTH

Readers will be familiar with the landcover photographs that are the basis for digital online maps. Beyond navigation, these satellite images can be processed to provide, for example, vegetation photosynthetic health trends. GIS maps presenting satellite image–derived raster data sets of biophysical parameters are useful for telling stories of threatened species habitat health. Farmers are now very familiar with the uncrewed aerial vehicle (UAV) images at centimeter spatial accuracy that provide farm management intervention, for example, where the agricultural inputs are needed across the farm, saving resources, money, and time.

Agronomists, foresters, and ecologists routinely use vegetation response rates of vegetation to understand the health trends of natural and agricultural resources, whether it is based on regular periods, seasons, climate modes (El Nino Southern Oscillation index, for example), or, depending on the satellite image archive, climate scale. This GIS *essay of place* uses the habitat area under scrutiny as the character and the temporal trends as the unfolding plot—does the GIS indicate a drying trend story? Dieback? Species change? These are essential questions that the GIS helps inform habitat managers and environmental policymakers alike. It's a story that supports ecological decisions.

A SATISFACTORY ENDING

Every story has an ending, and a story's success depends on how such an end resolves the narrative to a reader's satisfaction. Narrative maps similarly need to end well, and map design helps that process of narrative closure; everything in its place and a complete story. Field (2014, 99) is clear: "storytelling is the very essence of good map-making, and good cartographers have forever been successfully telling stories."

The language of cartography is a graphics structure to the meaning of a narrative and to convey a sense of place and concentrate a map reader's attention to critical spatial relationships. The GIS map-making craft elements are essential to a map becoming a story. The attention to narrative skill is analogous to the idea that anyone with a digital camera can take a picture, but can they make a picture that imparts impact? The GIS practitioner needs to ask, has the created map made a story? Indeed, Field reminds us that while the ease of GIS technology use allows for adroit map production, the cartographic framework and careful attention to craft and accompanying critical thinking is necessary to avoid ill-conceived and poorly constructed narratives. The master map-maker Gerardus Mercator knew maps could be compelling storytelling.

GIS map presentation is another vehicle for the nonfiction narrative—yet another way to tell stories that matter, another way to have an impact. An exemplar of map utility is, for example, the London Underground rail *tube* map. The map is held up as a carefully designed abstraction of the railway network such that travelers can make sense of the transport complexity. But that great map is information—it isn't yet a story. For a map to be story it needs to be crafted with elements analogous to the craft of creative writing.

Beyond being useful, Colloff (2020) believes that maps are a knowledge synthesis at the intersection of drawing and painting with writing and narrative. Even so, GIS craft must be skillfully applied to generate a mapping capacity to move a human being to change their mind like other communication devices of the arts: prose, poetry, and painting. The language of GIS is the cartographic conventions of symbols, labels, color, graphical abstraction, explanatory text, and the careful arrangement of all those elements, potentially providing the narrative. Beyond mere description, strategic stories have the power to change society's direction, to have an impact, to be a *net good*. The reader will bring to mind many novels, many poems that have changed minds. Has a map (GIS-enabled or not) that sort of power? One answer might lie in understanding literature scholar Brian Elliott's explanation (Colloff 2020, 20) of the stages through which a non-Indigenous Australian might travel to acquire an ecological awareness. The evaluation question progresses through five levels: is the landscape:

1. Merely topographical—what does the place look like?
2. Detailed and ecological—how does life arrange itself there?
3. Moral—how does the place influence people?
4. Indicating how do people make their mark on this place?
5. Holding subtler enquiries: what spiritual and emotional qualities emerge—do people develop here? What poetry emerges?

There is little argument against the notion that GIS helps map users gauge the landscape at points one and two above. The process of converting a two-dimensional map of symbols (perhaps mere semiotics of the landscape's story) to a realistic representation of the landscape takes satellite image base maps and modern three-dimensional GIS representations. These capabilities help people achieve the ecological awareness Elliott envisages. The spatial storytelling power of GIS spatial analysis, and other new knowledge-forming GIS utilities, can, to some extent, provide the third and fourth points above. Perhaps (to proffer an opinion) the GIS informs the appropriation of environmental consciousness levels, at least, providing the motivating information to want to acquire such knowledge personally. Suppose you agree that some of the GIS maps outlined in the vignettes discussed in the chapter approach the fifth point noted above. In that case, we might agree that GIS impacts society.

The advice of a well-known Australian playwright and artistic director Wesley Enoch (Schultz 2014, 2) is pertinent to the impact of culture: "do not ever underestimate that there are sometimes cultural solutions to intractable problems. When the law, economics and other systems fail, cultural and creative activities can work." The claim is bold, but the evidence of experience exists—GIS as a creative activity, and like other creative artifacts, GIS can move societal mountains!

REFERENCES

Alves, Daniel, and Ana Queiroz. "Exploring Literary Landscapes: From Texts to Spatiotemporal Analysis through Collaborative Work and GIS." *International Journal of Humanities and Arts Computing* 9 (2015): 57–73. https://doi.org/10.3366/ijhac.2015.0138.

Anthamatten, Peter. *How to Make Maps: An Introduction to Theory and Practice of Cartography*. Routledge, 2021.

Arthur, Bill, and Frances Morphy. *Macquarie Atlas of Indigenous Australia*. 2nd ed. Sydney: Macquarie Dictionary Publishers, 2019.

Brewer, Cynthia A. *Designing Better Maps: A Guide for GIS Users*. ESRI Press, 2005.

Burroway, Janet. *Imaginative Writing: The Elements of Craft*. 4th ed. Pearson, 2015.

Caquard, Sébastien, and William Cartwright. "Narrative Cartography: From Mapping Stories to the Narrative of Maps and Mapping." *The Cartographic Journal* 51, no. 2 (2014): 101–106. https://doi.org/10.1179/0008704114Z.000000000130.

Colloff, Matthew. *Landscapes of Our Hearts: Reconciling People and Environment*. Port Melbourne: Thames & Hudson, 2020.

Crane, Nicholas. *Mercator: The Man Who Mapped the Planet*. London: Weidenfeld and Nicholson, 2002.

El-Hadi, Nehal. "Writing Geography." *The Places Journal* (2020). Accessed 27th February 2020. https://placesjournal.org/reading-list/writing-geography.

Field, Kenneth. "Editorial: The Stories Maps Tell." *The Cartographic Journal* 51, no. 2 (2014): 99–100.

Garrard, Greg. *Ecocriticism*. Edited by John Drakakis.*The New Critical Idiom*. Abingdon and New York: Routledge, 2012.

Grayling, A C. *The History of Philosophy*. Viking, 2019.

Hughes, Christopher. "Bushfires in Australia—Statistics and Facts." Statista. (2021). Accessed 02/02/2021. https://www.statista.com.

ICSM. "Cartographic Considerations." Intergovernmental Committee on Surveying and Mapping. (2020). Accessed 6th June 2020. https://www.icsm.gov.au.

Konecny, Gottfried. *Geoinformation: Remote Sensing, Photogrammetry and Geographic Information Systems*. London and New York: Taylor & Francis, 2003.

Law, Michael, and Amy Collins. *Getting to Know ArcGIS Pro 2.6*. Redlands California: ESRI Press, 2020.

Longley, Paul A., Michael F. Goodchild, David J. Maguire, and David W. Rhind. *Geographic Information Science and Systems*. Wiley, 2015.

Lovelock, James. *Gaia: A New Look at Life on Earth*. Oxford University Press, 1979.

Mayhew, Susan. *A Dictionary of Geography*. 2nd ed: Oxford University Press, 1997.

Neale, Margo, and Lynne Kelly. *Songlines: The Power and Promise*. Edited by Margo Neale. *First Knowledges*. Port Melbourne: Thames and Hudson, 2020.

Olson, Randy. *Houston, We Have a Narrative: Why Science Needs Story*. Chicago: The University of Chicago Press, 2015.

Pearce, Margaret Wickens. "Framing the Days: Place and Narrative in Cartography." *Cartography and Geographic Information Science* 35, no. 1 (2008): 17–32. https://doi.org/10.1559/152304008783475661.

Perl, Sondra, and Mimi Schwartz. *Writing True: The Art and Craft of Creative Nonfiction*. 2nd ed. Boston: Wadsworth Cengage Learning, 2014.
Pickles, Thomas. *Map Reading*. London: J.M. Dent and Sons, 1960.
Reuschel, Anne-Kathrin Weber, Barbara Piatti, and Lorenz Hurni. "Data-Driven Expansion of Dense Regions: A Cartographic Approach in Literary Geography." *The Cartographic Journal* 51, no. 2 (2014): 123–40. https://doi.org/10.1179/1743277414Y.0000000077.
Ridanpaa, Juha. "Fact and Fiction: Metafictive Geography and Literary GIS." *Literary Geographies* 4, no. 2 (2018): 141–45.
Rigby, Colin Wayne, Alan Rosen, Helen Louise Berry, and Craig Richard Hart. "If the Land's Sick, We're Sick: The Impact of Prolonged Drought on the Social and Emotional Well-Being of Aboriginal Communities in Rural New South Wales." *The Australian Journal of Rural Health* 19 (2011): 249–54.
Ryan, Lyndall, William Pascoe, Jennifer Debenham, Stephanie Gilbert, Jonathan Richards, Robyn Smith, Chris Owen, Robert J Anders, Mark Brown, Daniel Price, Jack Newley, and Kaine Usher. "Colonial Frontier Massacres in Australia 1788–1872." Centre for 21st Century Humanities University of Newcastle, funded by ARC. (2017). DP 140100399. Accessed 30/10/2019. https://c21ch.newcastle.edu.au/colonialmassacres/map.php.
Schultz, David M. *Eloquent Science: A Practical Guide to Becoming a Better Writer, Speaker and Atmospheric Scientist*. Boston: American Meteorological Society, 2009.
Schultz, Julianne, ed. *Griffith Review 44: Cultural Solutions*. Edited by Julianne Schultz. Vol. 44: Griffith University, 2014.
———, ed. *Griffith Review 63: Writing the Country*. Edited by Julianne Schultz. Vol. 63: Griffith University, 2018.
Sui, Daniel. "Emerging GIS Themes and the Six Senses of the New Mind: Is GIS Becoming a Liberation Technology?" *Annals of GIS* 21, no. 1 (2015): 1–13. https://doi.org/10.1080/19475683.2014.992958.
Sword, Helen. *Air & Light & Time & Space*. Cambridge: Harvard University Press, 2017.
Taylor, Joanna E., Christopher E. Donaldson, Ian N. Gregory, and James O. Butler. "Mapping Digitally, Mapping Deep: Exploring Digital Literary Geographies." *Literary Geographies* 4, no. 1 (2018): 10–19.
Thomas, Leah. "Cartographic and Literary Intersections: Digital Literary Cartographies, Digital Humanities, and Libraries and Archives." *Journal of Map & Geography Libraries* 9, no. 3 (2013): 335–49. https://doi.org/10.1080/15420353.2013.823901.
UNEP. "Making Peace with Nature: A Scientific Blueprint to Tackle the Climate, Biodiversity and Pollution Emergencies." United Nations Environment Programme, Nairobi, 2021.
https://www.unep.org/resources/making-peace-nature.

Chapter Eight

Geomedia as a Pedagogical Tool
Toward Sustainability Competence

Michael John Long

One of the broad questions identified by this book, which is of interest to this chapter, is how have ecocinema and digital maps helped in the pursuit of critical thinking and reflective culture? In response, this chapter focuses on the usefulness of geomedia projects, here defined as "interactive technologies using multimedia to present the spatial information in the form of digital maps and georeferenced photos, videos and texts" (Motivate and Attract Students to Science n.d.) which include environmental documentaries as teaching tools for educators within liberal arts programs who seek to develop student skills and competencies related to sustainability.

    This chapter contends that a part of the role of educators who teach in the liberal arts should be to focus on sustainability competencies because of the pervasiveness of environmental issues. It has become integral that educators help to prepare students to enter the world and workforce, so that they are ready for life and work in relation to a rapidly changing climate and, as such, a rapidly changing world. This preparation is not restricted to teaching within the environmental humanities or toward work that includes environmental themes but applies to all programs within liberal arts because of the scale and reach of environmental concerns. One example of the need to pivot to teaching novel essential skills in response to a changing world was the emergence of digital technologies and the importance of preparing students for life and work that would come to include digital skills and competencies (UN 2019 chap. 2). As geomedia and education scholar Josef Strobl noted, "digital literacy is generally accepted as the passport into the information society, and from a GISociety perspective a certain set of geospatial literacy elements will be required" (Strobl 2014, 3). Geomedia is also such a pervasive force in daily life that teaching its intricacies to students is essential to helping them become literate citizens who can participate mindfully in society. Moreover, as this chapter explores, geomedia tools can be used to also teach sustainability competence.

The United Nations (UN) Sustainable Development Goals (SDGs) under SDG 4, Quality Education, aims for all learners to develop the skills and knowledge necessary to promote and achieve sustainable development. As this chapter highlights, subsequent work being conducted alongside those SDG 4 goals seeks to reorient education toward sustainable development. One of the ways to achieve that reorienting is to follow the work set out in the Education for Sustainable Development (ESD) framework, developed by the United Nations Educational, Scientific and Cultural Organization (UNESCO), which specifically "promotes competencies like critical thinking, imagining future scenarios and making decisions in a collaborative way" (UN 2014). The development of sustainability competencies is work taken up by scholars and departments within institutions of higher learning that have a specific focus on sustainability education. And yet, there is overlap between sustainability competencies and the soft skills and human skills that educators in liberal arts programs around the world focus on teaching that is worth exploring. Moreover, the need to incorporate sustainability into the strategic plans of many colleges and universities is becoming more evident. As an example, George Brown College of Applied Arts and Technology in Toronto, Ontario, Canada, in its Strategy 2022 and Vision 2030 plans, include both human skills and sustainability as fundamentally necessary for preparing students with future skills and behaviors to help "anticipate, absorb and manage change" (George Brown College n.d.).

This chapter further contends that geomedia and ecocinema can contribute to teaching sustainability competencies as pedagogical tools. In this chapter, ecocinema refers to didactic, issue-based environmental documentaries, or eco-docs, along with video essays in the form of short digital documentaries (e.g., VOX, Verge, Crash Course, TED). In this chapter, geomedia refers to projects which include eco-docs, alongside location-based data and interactive or digital maps, which have environmental issues at the core. In order to showcase how geomedia projects are helpful for teaching sustainability competencies, this chapter will conduct a brief case study using a geomedia project, the Anthropocene Education Project (AEP), which is a collaboration between award-winning filmmakers and nationally recognized educators. In concluding, this chapter will begin to lay the groundwork for educators to assemble "make-shift" geomedia projects in order to support the teaching of sustainability competencies.

REORIENTING EDUCATION

This section adopts the work conducted by academics who focus on reorienting education through the teaching of sustainability competencies. To start, this section relies on the UNESCO Chair in Reorienting Education Towards Sustainability, Charles Hopkins, and in particular, the chair's use of the ESD framework, which specifically encourages educators to teach competencies like critical thinking, futures thinking, and collaborative thinking. This section then moves to the work of sustainability and education scholar Arnim Weik, who has developed a comprehensive framework for defining and operationalizing sustainability competencies. In doing so, this section contends that reorienting education toward sustainability competencies is not only beneficial but is closely related to the work already being conducted by educators who teach human skills and soft skills in liberal arts programs.

The United Nations Sustainable Development Goals

The UN SDGs are an aspirational set of blueprints aimed at achieving a sustainable future by addressing issues of poverty, inequality, climate change, peace, and education, to name only a few. In September 2020, Malala Yousafzai, the UN Messenger of Peace and Nobel Laureate, delivered virtual remarks at the SDGs Moment event, which ran alongside the 75th Session of the United Nations General Assembly. The SDGs Moment aimed to kickstart the Decade of Action to mobilize governments, civil society, businesses, and individuals to ensure the delivery of the seventeen Global Goals which were set in 2015 and are sought to be achieved by 2030.

In a prerecorded people's address, Yousafzai requested that world leaders "set the norms of a new era—a sustainable, healthy, educated and equitable era" (United Nations 2020). An important thread that was woven through Yousafzai's speech was the topic of education. She acknowledged that while the COVID-19 pandemic has been a "striking setback to our collective goals, it cannot be an excuse," and "on education alone, 20 million more girls may never go back to the classroom when this crisis ends" (UN 2020). At the same event, UN Secretary-General Antonio Guterres reiterated that sentiment by noting that the world has been "shaken to the core" by "the pandemic which has pushed us towards the worst recession in decades, with terrible consequences for the most vulnerable—societies and citizens are reeling from widespread disruption, after many years of progress, poverty and hunger are on the rise, [and] children are suffering from a lack of schooling" (UN 2020b). It is evident that education is a notable agenda item for those working on the SDGs moving into the Decade of Action.

There are seventeen SDGs. SDG 4, Quality Education, is essential because it underpins the other goals and highlights literacy as key to developing sustainable societies. Overall, SDG 4 aims to provide children and youth with quality and accessible education and further learning opportunities, and by 2030 "ensure that all learners acquire the knowledge and skills needed to promote sustainable development" (UN 2020c). This is important because education is a key contributor to sustainable development, nation building, and peace building. SDG 4 is a goal set for the world at large and contains seven targets to help meet the overall goal, and twelve indicators from which to track progress toward each target. The seven targets cover topics like equal access, increased acceptance, elimination of discrimination, universal literacy, inclusivity and safety. And yet, there has been particular attention paid to target 4.7:

> By 2030, ensure that all learners acquire the knowledge and skills needed to promote sustainable development, including, among others, through education for sustainable development and sustainable lifestyles, human rights, gender equality, promotion of a culture of peace and non-violence, global citizenship and appreciation of cultural diversity and of culture's contribution to sustainable development. (UN 2020c)

An example of one way in which target 4.7 has been highlighted is through the development of Global Citizenship Education (GCED) as one of UNESCO's responses to the human rights violations, inequality, and poverty that occur in an interconnected world, and which threaten peace and sustainability.

Moreover, indicator 4.7.1 states that the "extent to which (i) global citizenship education and (ii) education for sustainable development, including gender equality and human rights, are mainstreamed at all levels in (a) national education policies; (b) curricula; (c) teacher education; and (d) student assessment" (UN 2020c). This indicator, which offers areas to track within the journey toward mainstreaming education for sustainable development goals, may be an important recommendation for how to also implement those goals. For educators working in liberal arts programs, subsections (ii) (a) to (d) may be notably helpful.

A UN affiliate body which has taken up the work on SDG 4.7 and 4.7.1 is the UNESCO Chair in Reorienting Education Towards Sustainability. The chair, which is housed at York University, Toronto, Ontario, Canada, was established to work on education for sustainable development and to create guidelines for reorienting teacher education to address sustainability. An important aspect of the work conducted by the chair is the idea that educating people about sustainable development is not enough, but it is also integral to reconsider education systems entirely.

In making this argument, the chair relies on the framework set out in the ESD program, which was a UN initiative, delivered by UNESCO, and promoted by the Decade of Education for Sustainable Development 2005–2014. ESD is a way of engaging with education systems, in order to revisit the idea of what quality education entails toward a more sustainable future. The 2012 ESD Sourcebook, which was a manual published for beginning the process of combining education and sustainability, notes four major areas of intervention through which to support sustainable development in education systems: 1) access to and retention in a quality basic education; 2) reorienting education systems to address sustainability; 3) to build public awareness and understanding of sustainable development; and 4) to provide training to all in both the private and public sectors to promote sustainability at home and in the workplace (UNESCO Chair in Reorienting Education towards Sustainability n.d.). The area of intervention which is of immediate interest for this section is number two.

The use of the ESD framework is helpful in implementing the SDG 4.7 and 4.7.1 goals, and ultimately reorienting education systems to transform society and "help[ing] people develop knowledge, skills, values and behaviours needed for sustainable development" (UN 2020c). The ESD framework outlines the necessary elements that need to be addressed in order to achieve these goals:

> ... including key sustainable development issues into teaching and learning; for example, climate change, disaster risk reduction, biodiversity, poverty reduction, and sustainable consumption ... It also requires participatory teaching and learning methods that motivate and empower learners to change their behaviours and take action for sustainable development. Education for Sustainable Development consequently promotes competencies like critical thinking, imagining future scenarios and making decisions in a collaborative way. (UN 2014)

Moreover, ESD has continued to gain increased importance because of its contributions to advancements on SDG 4. In November 2019, the 40th Session of UNESCO General Conference adopted a new framework on ESD, "Education for Sustainable Development: Toward Achieving the SDGs," or "ESD for 2030" for short. ESD for 2030 "aims to scale up action from the United Nations Decade of Education for Sustainable Development (2005–2014) and the Global Action Programme (GAP) on ESD (2015–2019)" (UN3 2020) according to the UNESCO website, and which will launch at UNESCO World Conference on Education for Sustainable Development in Berlin on May 17–19, 2021. It is clear that this is an important line of work for education in sustainable development for the UN and its affiliated bodies.

Sustainability Competence

The goal of reorienting education toward sustainable development and doing so through the ESD framework (in order to " . . . promote competencies like critical thinking, imagining future scenarios and making decisions in a collaborative way" [UN 2014]) is work that is being conducted by dedicated sustainability programs in institutions of higher learning around the world. Sustainability and education scholar Arnim Weik, professor in the School of Sustainability, Arizona State University (ASU), is a leader in the work on amalgamating sustainability and competencies. Weik et al. write that "competencies have received increasing attention as critical reference points for the development of curricula and courses" (Wiek et al. 2015, 242). Weik et al. define "competence as 'a functionally linked complex of knowledge, skills, and attitudes that enable successful task performance and problem solving'; applied to competencies in *sustainability*, these are 'complexes of knowledge, skills, and attitudes that enable successful task performance and problem solving with respect to real-world sustainability problems, challenges, and opportunities'" (Wiek et al. 2015, 242).

The competencies themselves are outlined by Weik et al. who completed a broad literature review of sustainability competencies in higher education and synthesized that work into five key categories described below (Weik, Withycombe, and Redman 2011, 203–218). It is important to note here that this chapter is not a review of the competencies nor is it an examination of how to operationalize the competencies because that work has already been expertly conducted. Instead, as a reminder to the reader, this chapter considers how geomedia and ecocinema might be helpful in successfully teaching at least a few of the five key competencies to which they are particularly suited. And so, it will be helpful here to define the competencies to be able to return to them in subsequent sections. The following definitions have been adopted by the ASU School of Sustainability, and were developed from the work of Weik et al. It is also important to make two further notes before moving on to the definitions. The first is that the sixth competency not defined below is the meaningful integration of the mentioned five. The second note is that a key aspect of competencies is that they include the issue-based or topical-based knowledge within a particular context.

1. Systems Thinking Competence is defined as being "able to analyze sustainability problems cutting across different domains (or sectors) and scales (i.e., from local to global), thereby applying systems concepts including systems ontologies, cause-effect structures, cascading effects, inertia, feedback loops, structuration, etc." (Weik, Withycombe, and Redman 2011, 203–218). As adopted from the work of Weik et al., a

key competence inherent here is problem solving from a systemic perspective, for example, the ability to identify how different professional activities help to solve or mitigate sustainability issues.
2. Futures Thinking (or Anticipatory) Competence is defined as being "able to anticipate how sustainability problems might evolve or occur over time (scenarios), considering inertia, path dependencies, and triggering events; as well as creating and crafting sustainable and desirable future visions, considering evidence-supported alternative development pathways" (Arizona State University n.d., 3). As adopted from the work of Weik et al., a key competence inherent here is, again, problem solving but from an anticipatory perspective, for example, the ability to identify how professional activities evolve and contribute to solving or mitigating sustainability issues.
3. Values Thinking (or Normative) Competence is defined as being "able to specify, compare, apply, reconcile, and negotiate sustainability values, principles, goals, and targets, informed by concepts of justice, equity, responsibility, etc., in various processes, including visioning, assessment, and evaluation" (Arizona State University n.d., 4). As adopted from the work of Weik et al., a few key competencies inherent here are empathy, ethics, justice, and fairness, for example, the ability to comprehend and implement notions of integrity, harm, damage, stewardship, conservation, and more in the execution of professional activities.
4. Strategic Thinking (or Action-Oriented) Competence is defined as being "able to design and implement systems interventions, transformational actions, and transition strategies toward sustainability, accounting for unintended consequences and cascading effects" (Arizona State University n.d., 5). As adopted from the work of Weik et al., a key competence inherent here is critical thinking, for example, the ability to exercise thoughtful decision-making, designing plans, identifying barriers to plans, and intentional actions toward transformational change.
5. Collaboration (or Interpersonal) Competence is defined as being "able to motivate, enable, and facilitate collaboration towards sustainability" (Arizona State University n.d., 6). As adopted from the work of Weik et al., a few key competencies inherent here are verbal, written and electronic (e-) communication, project management, and cross-cultural collaboration, for example, the ability to conduct work with diverse groups of stakeholders.

Weik et al. note that "sustainability programmes in higher education institutions are supposed to convey these competencies in sustainability and enable graduates to make contributions to resolving challenging societal problems and building a sustainable future" (Wiek et al. 2015, 242). The authors also note that "large-scale educational transformation is needed to equip a new generation of professionals (not only *sustainability* professionals!) to address sustainability challenges through problem-solving approaches that integrate systems thinking, structured anticipation, value-laden deliberation, evidence-supported strategies, and strong collaboration across government, businesses and civil society" (Wiek et al. 2015, 241). For educators who operate within a wide range of fields in the liberal arts and humanities, teaching these competencies can be used to achieve issues-based teaching goals and sustainability teaching goals, simultaneously. In fact, sustainability competencies may likely be a familiar concept for educators within the liberal arts and humanities because they overlap with some of the skills which are taught in order to ensure that graduates can contribute well to the world and workforce. This section contends that the five key sustainability competencies framework is applicable within the liberal arts and humanities for use by educators who seek to include sustainability into their teaching goals. Moreover, the inclusion of sustainability competencies is an increasingly common goal for institutions as liberal arts programs seek to amalgamate sustainability and human skills into their future goals—George Brown College is one example among many (George Brown College n.d.).

Human Skills and Soft Skills

The sustainability competencies framework could be beneficial for educators in liberal arts and humanities programs who are already particularly attuned to teaching human skills and soft skills. Human skills are the skills needed to relate with one another like communication (see Collaboration or Interpersonal Competence) and empathy (see Values Thinking or Normative Competence). Soft skills are the skills that relate to personality traits, habits, and behaviors like problem solving (see Futures Thinking or Anticipatory Competence), collaboration, and integrity. These skills are typically taught alongside issue-based knowledge with broad-based personal and/or societal importance, and as such increasing awareness of the society and culture in which people live and work.

It is the above types of skills that employers seek. According to a 2019 LinkedIn survey, 92 percent of global talent professionals, such as human resource experts, business leaders, and talent acquisition said that "soft skills matter as much or more than hard skills" (LinkedIn 2019), while 80 percent said that soft skills are "increasingly important to company success" (LinkedIn 2019). In the survey, the soft skills noted to be in current demand are creativity, persuasion, collaboration, adaptability, and time management. Those in-demand soft skills are some of the same noted in another recent survey which focused on careers within Canada's environmental sector. In 2018, the Environmental Careers Organization of Canada (ECO Canada) surveyed employers and employees within the fields of environmental protection, resource management, and sustainability, and determined that although technical skills (policy and legislation, industry knowledge, sustainability, site assessment and reclamation, and climate change) are essential for initial hiring, it is the soft skills of communication, collaboration, project management, report writing, and attitude and& professionalism that are sought after by employers (ECO Canada 2019).

PEDAGOGICAL TOOLS FOR TEACHING SUSTAINABILITY COMPETENCE

This section contends that geomedia and ecocinema can be used as pedagogical tools for educators to help achieve the teaching of sustainability competencies. It is first necessary to explore the various ways that geomedia and ecocinema are broadly interlinked, and to note the version of the form which has been particular useful in my teaching practice as a Contract Faculty Professor of Geography and Globalization, in the School of Liberal Arts & Sciences, at George Brown College of Applied Arts and Technology, in Toronto, Ontario, Canada, and has developed in relation to work as a programming team member at Planet in Focus International Environmental Film Festival (PIF) and the associate programmer at Water Docs Film Festival.

Where Geomedia and Ecocinema Meet

This section focuses on the intersection of environmental documentary, sometimes called eco-docs, as well as digital documentaries in the form of video essays (e.g., VOX, Verge, Crash Course, TED), and geomedia projects which utilize location-based data, digital maps, academic research, and eco-docs. However, it is important to acknowledge that there are many types of documentaries that are useful as teaching tools. Along with eco-docs, the field of ecocinema also includes fiction and nonfiction films which explore environmental themes. Digital documentaries, sometimes referred to as doc-media, includes interactive documentaries (i-docs), web-docs, and multilinear documentaries. There are also database documentaries, sometimes referred to as living documentaries, which have infinite content potential (Terry 2020, 149). And, of course, there is overlap between the types listed above, and with the types on which this section focuses, eco-docs. Similarly, there are many types of geomedia projects that are useful for teaching. As mentioned earlier, geomedia is a relatively recent and expanding term that can refer to "interactive technologies using multimedia to present the spatial information in the form of digital maps and georeferenced photos, videos and texts" (Motivate and Attract Students to Science n.d.). Imagine anything from geo-tagged pictures or social media posts to vehicle navigation systems, to restaurant reviews, to academic research, and far beyond.

Geomedia projects can be particularly useful for sustainability teaching goals especially when they unite the geomedia genre, including location-based data and interactive or digital maps, with the digital documentary genre, including web-based eco-docs. This is because eco-docs are often, although not exclusively, didactic, issue-based, and sociopolitical in nature. Moreover, digital documentaries are useful here for their educational content. And this is rather than, for example, mountaineering films (*Meru* 2015, *Free Solo* 2018) or feature fiction films with environmental themes (*Twister* 1996, *Night Moves* 2013). The reason my teaching practice has focused on geomedia is because information is increasingly packaged and shared by way of maps, visualizations, images, surveys, documents, and more. As the editors of the book *Learning and Teaching with Geomedia* note, "our lives are thoroughly geomedialized" (Gryl et al. 2014, viii) with students having increasing access to social media, web-based searchable data, and the cloud, and all of which can be entered through relatively inexpensive smartphones, tablets, and laptops, not to mention popular virtual reality and augmented reality headsets.

One example of the meeting of geomedia and eco-docs is the Geo-Doc. In his book, *The Geo-Doc: Geomedia, Documentary Film, and Social Change* (2020), Mark Terry explores the potential of merging digital documentaries (i-docs, web-docs, multilinear docs, database docs) with geomedia in order to effect positive social change on the part of high-level changemakers, such as at the UN. As exemplified with his project, the *Youth Climate Report* (Terry, UNCCS 2020), Terry determines that digital documentaries united with geomedia can help bridge the communication gap between scientists and policymakers, in order to convey messages in easily accessible ways. In January 2021, the YCR won an Honorable Mention at the UN's SDG Action Awards, recognizing the project's ability to mobilize youth, give voice to under-represented communities, and bridge the science-policy gap (UN 2021d). Terry has also noted, importantly for this chapter's aim, that his version of the form, the Geo-Doc, can and has been used as a pedagogical tool because it offers a wealthy database of issue-based information, and opportunities for enhanced experiential education because students are able to create their own Geo-Docs, as outlined through a step-by-step guide for educators and leaners at the end of his book.

Case Study on Teaching Sustainability Competencies

Another version of the form which can be useful for teaching sustainability competencies is the Anthropocene Education Program (AEP) collaboration between The Anthropocene Project (TAP) and the Royal Canadian Geographical Society (RCGS) (Canadian Geographic n.d.). This particular geomedia project includes environmental documentary, short films, audio clips, augmented reality, 360 virtual reality, interactive gigapixel photos, and issued-based educational material, all of which is freely available through Canadian Geographic Education. The AEP gives educators access to these tools to teach students about the impact that humans have on the Earth. TAP was created by the award-winning documentarians Edward Burtynsky, Jennifer Baichwal, and Nick de Pencier (*Manufactured Landscapes* 2006, *Watermark* 2013, *Anthropocene: The Human Epoch* 2019) to create "an opportunity for students to learn the history and science behind the Anthropocene and understand just how much humans are changing Earth's natural systems" (Canadian Geographic n.d.). The project is informed by the Anthropocene Working Group (AWG), a scientific committee tasked by the Subcommission on Quaternary Stratigraphy in 2009, to gather and review evidence, and recommend whether the Anthropocene should replace the Holocene as the current geologic time period. The term itself, Anthropocene, was introduced and popularized by AWG members, Eugene Stoermer and Paul Crutzen, in order to suggest that humans are living in a new geologic epoch characterized by the scale of our planetary change (Steffen et al. 2011, 842–867).

The AEP is an astonishingly entertaining and educational geomedia project. There are various ways to access the project, the first of which is the free, on-loan classroom kit that includes virtual reality headsets, augmented reality experiences, tablets, and books. The second is the free online resources, which include a teacher's guide, fact cards, and supporting short films and 360 virtual reality films for use on computer, tablet, or phone screens. One of the key online resources are the Interactive Gigapixel Photos, which are original mosaic photographs taken by award winning artist, Edward Burtynsky, which are combined to create billion-pixel, hyperdetailed, large-scale explorable images. Using a computer, tablet, or phone, users can navigate around the images, zoom in and out, and explore embedded visual triggers that reveal short films, audio clips, or 360 videos. When explored together, the triggers reveal a story about the type of human impact on the natural world, which is rooted in the broad categories of impact outlined by the AWGs research: terraforming, techno-fossils, anthroturbation, climate change, extinction, and extraction. There is also issue-based educational material which consists of fourteen lesson plans developed in association with the Canadian Geography Learning Framework, aimed at grades four to twelve. These lesson plans are a rich database of information that can be used to teach topical knowledge, but it is important to note that, as with other geomedia projects, it is often necessary that educators supplement some of the issue-based knowledge with instruction in order to reach the desired level of novice, intermediate, or advanced teaching. By adding to the educational materials, educators can offer students a bird's-eye view of any global environmental issue that intersects with the concept of the Anthropocene.

The AEP is well suited to help educators teach at least a few of the sustainability competencies highlighted in the section above (Sustainability Competence). Sustainability Competence 1, Systems Thinking, has at its core problem solving from a systemic perspective as an essential skill. One way that a geomedia project like the AEP can help teach this competency is by showcasing the relationships between subject matter. A benefit of the Anthropocene, as both a popularized concept and a geomedia project, is that it offers an overview of the various categories of human impact on the Earth and how those categories relate to each other, and ultimately, how they impact earth system functioning (Steffen, Crutzen, and McNeill 2007). In order to attend to environmental and sustainability issues, it is integral to understand the complexity and interconnectedness of those issues.

Sustainability Competence 3, Values (or Normative) Thinking, has at its core empathy, ethics, justice, and fairness as essential skills. As mentioned above, the AEP can offer opportunities for geo-relational study, and allow users to connect with the experiences and perspectives of other people from around the world. One way that the AEP can achieve this is through its utilizing its eco-doc component, whether in long or short form films adopted from the feature, *Anthropocene: The Human Epoch* (2019), which preceded the project. This eco-doc has environmental issues at the centre and uses storytelling techniques to invoke concepts of empathy, ethics, justice, and fairness that can be helpful for students in the process of implementing notions of integrity, stewardship, and conservation into their work. As Terry notes "documentaries that tell stories not only of our home, but our place in it have the potential to engage the viewer emotionally and compel them to action if the message is how we are acting as homewreckers" (Terry 2020, 90). The AEP, as a geomedia project which incorporates a familiar eco-doc format and an unconventional multimedia narrative, is providing, to use Terry's words, "an altogether new way to show humans the world they so adversely impacted and to encourage them to act to create progressive environmental and behavioural change on a global basis" (Terry 2020, 97).

Sustainability Competence 5, Collaboration (or Interpersonal) Thinking, has at its core verbal, written, and e-communication; project management; and cross-cultural collaboration as essential skills. There are numerous ways to use the AEP to teach collaboration, for example. One way is to show how the success of the project is reliant on its collaborative nature through its utilization of numerous mediums (film, AR, VR, interactive photos, text-based learning materials, etc.) and its numerous partners (AEP, RCGS, Canadian Geographic Education, funders, etc.). Moreover, educators can teach how the various mediums have their own inherent ways of reaching and communicating information to the user. For example, the way that the AEP communicates information is with highly produced and engaging audio and visual content that contextualizes the material for more accessible digestion and understanding.

THE "MAKE-SHIFT" GEOMEDIA TOOLKIT FOR EDUCATORS

Geomedia projects with the depth and production quality of the AEP and YCR are rare. However, other geomedia projects which can be helpful for teaching sustainability competencies are accessible on the internet—readers interested in exploring these further can reference the nonexhaustive list in the resource section below. Other stand-alone resources exist as well in the form of interactive maps and the research used to generate them, which are key elements of geomedia projects. This section contends that educators can create "make-shift" geomedia packages from existing and related multimedia resources available on the Internet to formulate geomedia teaching packages. In order to do so, there are a few important considerations. One is that to ensure the "make-shift" geomedia package is well-rounded, it should contain, at minimum, an interactive or digital map, a digital documentary, an accessible mainstream article, an in-depth academic journal, all of which is concentrated on a specific and related issue.

An example of a "make-shift" geomedia package could center around the field of cartography and helps to teach about map projections and projection inaccuracies. In developing this package, an educator could use the interactive map created by James Talmage and Damon Maneice, "The True Size Of . . . " (Talmage and Maneice n.d.), which allows for learner to select a country and drag the country to another part of the Mercator projection map, in order to visualize how that country's shape/size changes as it gets closer or further from the poles (Briney 2014). The educator could also utilize a digital documentary, perhaps in the form of a video essay available on Vox Media's YouTube channel, titled "All Maps Are Wrong. I Cut Open a Globe to Show Why" that uses audiovisual tools to show how our 3D spherical planet can be misrepresented when transferred for use in a flat 2D representation. This video essay provides a reinforcement of the interactive map lesson and offers an alternative communication method for the sharing of information (Harris

2016). The next set of geomedia tools used by the educator could be accessible mainstream media articles, such as "Africa Could Fit China and the U.S., with Room to Spare," published by *The Atlantic* which directly references the interactive map mentioned earlier (Mohammed 2015). Lastly, the educator could also use an academic journal or course textbook entry on the art and science map projections.

These "make-shift" geomedia projects discussed have been assembled over the years in order to enhance teaching of issue-based knowledge and sustainability competencies, simultaneously. The two examples below cover megacities and sea-level rise, but there are numerous options for other on subjects that range from invasive species to green roofs to global interconnected networks, and beyond. This chapter aims to encourage educators to begin to assemble "make-shift" geomedia packages to enhance their own teaching practice.

CONCLUSION

In the midst of a global health pandemic, as the integration of educational technology into pedagogy becomes increasingly salient, the use of geomedia may even be expandable and scalable. One of the ways that the COVID-19 pandemic has been challenging is due to the rapid transition to online teaching and learning as the primary mode of education for institutions around the world. In March 2020, as concerns rose over COVID-19, numerous Canadian postsecondary institutions transitioned, mid-semester, from in-person to online delivery. In May 2020, those colleges and universities announced plans for primarily online delivery for that upcoming fall semester and, at the time of writing, did so again for the upcoming winter 2021 and spring 2021 semesters. As education technologies continue to be implemented and will likely find a place for both in-person and online teaching and learning, geomedia may have further roles to play.

The line of thought above introduces another broad question asked by this book: how can transmedia projects incorporate environmental research and GIS platforms? A potentially productive expansion of the investigation in this chapter could be to determine if geomedia projects can benefit college or university campuses at large, beyond the classroom. A geomedia project that includes interactive or digital maps, issue-based information, and short documentaries could be created to highlight the sustainability efforts by the various departments on campus. This work could simultaneously be a bridge-building exercise and a cataloguing of the sustainability work conducted by seemingly disparate parts of the campus. The geomedia platform could be helpful in telling the story of campus sustainability efforts and identify gaps that need to be addressed. Perhaps the geomedia genre can be applied to both the classroom setting and toward campus sustainability goals at large.

REFERENCES

Arizona State University. ""Key Competencies in Sustainability." Teaching Resources. Accessed October 2020. https://static.sustainability.asu.edu/schoolMS/sites/4/2018/04/Key_Competencies_Overview_Final.pdf.

Briney, Amanda. "A Look at the Mercator Projection." *GIS Lounge* (2014). Accessed February 2021, https://www.gislounge.com/look-mercator-projection/.

Canadian Geographic. "Anthropocene Education Program." Accessed October 2020. https://anthropocene.canadiangeographic.ca/.

Environmental Careers Organization of Canada (ECO Canada). "Skills Essential for Success in the Environmental Sector." (2019). Accessed October 2020. https://eco.ca/new-reports/developing-skills-to-succeed-in-the-environmental-sector/.

George Brown College. "Imagining Possibilities: Vision 2030, Strategy 2022." Accessed October 2020. https://www.georgebrown.ca/sites/default/files/2020-05/Imagining%20Possibilities%20-%20Vision%202030%20Strategy%202022.pdf.

Gryl, Inga, Caroline Juneau-Sion, Eric Sanchez, John Lyon, and Thomas Jekel, eds. *Learning and Teaching with Geomedia.* Tyne: Cambridge Scholars Publishing, 2014.

Harris, Johnny. "All Maps Are Wrong. I Cut Open a Globe to Show Why." *Vox* (2016). Video, 6:00. https://www.vox.com/world/2016/12/2/13817712/map-projection-mercator-globe.

LinkedIn. "2019 Global Talent Trends Report: The 4 Trends Transforming Your Workplace." LinkedIn Talent Solutions (2019). Accessed October 2020. https://business.linkedin.com/talent-solutions/blog/trends-and-research/2019/global-recruiting-trends-2019.

Mohammed, Omar. "Africa Could Fit China and the U.S., with Room to Spare." Global. *The Atlantic*, August 18, 2015. https://www.theatlantic.com/international/archive/2015/08/mercator-projection-distortion-map-africa/401543/.

Motivate and Attract Students to Science. "General Description." Geo-media. Accessed October 2020. http://www.mass4education.eu/geo-media.

Steffen, Will, Jacques Grinevald, Paul Crutzen, and John McNeill. "The Anthropocene: Conceptual and Historical Perspectives." *Philosophical Transactions of the Royal Society* A 369 (2011): 842–67. https://doi.org/10.1098/rsta.2010.0327.

Steffen, Will, Paul J. Crutzen, and John R. McNeill. "The Anthropocene: Are Humans Now Overwhelming the Great Forces of Nature?" *Ambio* 36, no. 8 (2007). https://www.jstor.org/stable/25547826.

Strobl, Josef. "Technological Foundations for the GISociety." In *Learning and Teaching with Geomedia*, edited by Thomas Jekel, Eric Sanchez, Inga Gryl, Caroline Juneau-Sion, and John Lyon, 2–9. Tyne: Cambridge Scholars Publishing, 2014.

Talmage, James, and Damon Maneice. "The True Size Of." Accessed October 2020. https://thetruesize.com/.

Terry, Mark. *The Geo-Doc: Geomedia, Documentary Film, and Social Change.* Palgrave Macmillan, 2020.

Terry, Mark, and the United Nations Climate Change Secretariat (UNCCS). "*Youth Climate Report.*" Accessed October 2020. https://youthclimatereport.org/.

United Nations (UN) Conference on Trade and Development. "Building Digital Competencies to Benefit from Frontier Technologies." New York: United Nations Publications, 2019. https://unctad.org/system/files/official-document/dtlstict2019d3_en.pdf.

United Nations (UN) Educational, Scientific and Cultural Organization. "What Is ESD?" UNESCO, 2014. Accessed October 2020. http://www.unesco.org/new/en/unesco-world-conference-on-esd-2014/resources/what-is-esd/.

United Nations (UN). 2020a. "Sustainable Development Goals Are 'the Future' Malala Tells Major UN Event, Urging Countries to Get on Track." *UN News*, September 18. https://news.un.org/en/story/2020/09/1072782.

United Nations (UN). "Sustainable Development Goals: 4 Quality Education." UN, 2020b. Accessed October, 2020. https://www.un.org/sustainabledevelopment/education/.

United Nations (UN) Educational, Scientific and Cultural Organization. "Education for Sustainable Development." UN, 2020c. Accessed October 2020. https://en.unesco.org/themes/education-sustainable-development.

United Nations (UN) Sustainable Development Goals Action Awards. 2021. 'Honourable Mentions—Youth Climate Report." UN, 2020d. Accessed February 2021, https://sdgactionawards.org/youth-climate-report/.

Wiek, A., Bernstein, M., Foley, R., Cohen, M., Forrest, N., Kuzdas, C., Kay, B., & Withycombe Keeler, L. "Operationalising Competencies in Higher Education for Sustainable Development." In *Handbook of Higher Education for Sustainable Development*, edited by M. Barth, G Michelsen, M. Rieckmann, I. Thomas, 241–260. London: Routledge, 2015.

Weik, Arnim, Lauren Withycombe, and Charles L. Redman. "Key Competencies in Sustainability: A Reference Framework for Academic Program Development." *Sustainability Science* 6 (2011): 203–18. https://doi.org/10.1007/s11625-011-0132-6.

Chapter Nine

When Place Is Elsewhere

Pedagogy of Place for Planetary Health Education in a Digital Space

Netta Kornberg

In February 2020, the Planetary Health Film Lab took place at the Dahdaleh Institute for Global Health Research, York University, in Toronto, Canada. The Planetary Health Film Lab is a collaboration between the Dahdaleh Institute for Global Health Research, the Young Lives Research Laboratory, and the Youth Climate Report; research was supported by the Social Sciences and Humanities Research Council of Canada. The project, described below in detail, engaged young adults from different parts of the world to make short documentaries about the impacts of climate change on health in their home neighborhood, city, or country. The project's aims were both pedagogical and advocative, and it made use of multiple digital technologies.

Core to the project was geographic information system (GIS) mapping, a technology of place whose purpose in the project was, in part, to create "geographies of hope" (Pavlovskaya 2018) in which participants could see their passion for the environment and the challenges of climate change mirrored by the passion and struggle of other young adults. GIS can enable this largely through documentation, representation, and imagination. This aspiration for the participants' learning experience came to fruition only in part. Tension emerged between GIS as a data-based technology of place and the place-based learning which is typically central to a pedagogy of place.

According to Brett Hirsch, the initial interest of digital humanities in pedagogy, strong throughout the 1980s and 1990s, practically disappeared in the new millennium (Hirsch 2012). On the one hand, much is taken for granted about the appropriateness of GIS technologies for education. On the other, critical views on the role of technology in ecological education too often stop at critique. The issue of technology in education became urgent in the months after the Film Lab, which took place one month before the World Health Organization declared Coronavirus a pandemic (Ghebreyesus 2020). Less than three weeks after the Film Lab concluded, the world began to close off our physical spaces and went online at a perhaps unprecedented scale. In a sometimes clumsy rush, postsecondary education has been moved mostly to digital, and the future of adult education—within and outside postsecondary institutions—remains unknown as of the writing of this chapter.

I am an adult educator who never advocated for a central role of digital technologies in the learning experience. Yet digital technologies in education have become inevitable for a time, perhaps for a long time.

This period of digital learning coincides not only with the coronavirus pandemic, but also with climate change, a likely existential threat to the future of the human species.

These two global crises forefront that humans are part of the environment, and that human wellbeing is intertwined with planetary wellbeing. They forefront the need for human relationships with the planet to mature (McGregor 2020). And where we talk about maturation, we talk about education. It is more relevant than ever to foster rich educational approaches to planetary health, and many of those educational experiences are taking place in digital spaces, many of them formed through geographic information systems.

This chapter is written as a reckoning with the present and possibilities of place-based adult education, an affirmation of core values of education, and a commitment to planetary health.

ABOUT THE PROJECT

I co-organized the program with an editor of this volume. A host of mentors, presenters, and supporters delivered workshops and provided guidance and technical support. The workshop lasted five days. It involved seven participants from six countries: one each from Australia, Canada, Italy, India, and Colombia, and two from Ecuador. The participants were nineteen to twenty-four years old, the youngest having just started postsecondary school, the oldest having recently graduated and working or in a graduate program. Many were completing undergraduate degrees.

The call for applications for the Film Lab described a workshop, all costs covered, open to youth from anywhere in the world. At the end of the workshop, each participant would have completed a short documentary film about climate change and health in their home community. This was our call:

The Planetary Health Film Lab is an intensive program designed for youth who have a story to tell about climate change and health and want to do so through film.

During a week-long workshop at York University, Toronto, twelve international and domestic participants will learn to effectively tell stories that communicate data, research, and life experiences related to global and planetary health. The workshop teaches specific theories, techniques, and modes of social issue filmmaking and provides hands-on experience with new digital technologies and platforms.

During the program, participants produce documentary short films that will be featured on the websites of the United Nations Framework Convention on Climate Change, the Dahdaleh Institute for Global Health Research and the Youth Climate Report, influential platforms used as a resource by policy-makers. The films will directly contribute to progressive policy creation on a global scale.

The purpose of the lab was to foster a group of climate advocates through the skills and experience of filmmaking, science communication, and storytelling.

Geographic information systems (GIS) played a central role. The endpoint of the films was the Youth Climate Report, a preexisting project headed by Mark Terry. The Youth Climate Report consists of short films created by youth worldwide reporting on climate change. Those films are hosted on a GIS map, and the best films every year are presented at the United Nations Framework Convention on Climate Change conferences (COP). Youth Climate Report is a partner of the United Nations Climate Change Secretariat, and an emphasis is put on effective ways of communicating with policy-makers in this domain.

CHAPTER SUMMARY

The purpose of this chapter is to learn from the Planetary Health Film Lab, the methods we chose to use, the ways in which we succeeded, the shortcomings of the program, and the insights of the participants. This chapter will first define key terms before giving more detail about the Film Lab, followed by a reflection on the challenges and opportunities of place-based learning in a digital space. I will examine the role of GIS as the primary technology of place engaged in the workshop. I will then argue for three components of place-based pedagogy for planetary health education through digital technology: content knowledge, love, and storytelling.

PARTICIPANT VOICES

Throughout this paper, long quotes from Film Lab participants are included to contextualize and enrich. These quotes are from exit interviews conducted with each participant. Most participants' first language is not English, and that is reflected in the transcribed text. All participants were insightful, energetic, thoughtful, and brought their best to the program.

WHAT I TALK ABOUT WHEN I TALK ABOUT PLACE-BASED PLANETARY HEALTH EDUCATION

The following is a review of key terms and concepts which make up the arguments in this paper.

Adult Education

> It all started with the November floods in Venice. Before then I wasn't really thinking about climate change, I wasn't really thinking that I could do something about it with my filmmaking skills. And then the Film Lab came up and it was something really new. I had never done anything about planetary health or about the environment in my videos or short films. I wasn't really even interested to be honest, so it was a turning point because I started to think about it, and I started to think about it in terms of what I could do, using my passion which is filmmaking. (Participant 2020-03 2020)

In his critique of education in Africa, Charles Kabuga characterized its pedagogy, exported from Europe to Africa through colonialism, as one which "oppressed, silenced, and domesticated the learner" (Kabuga 1984, 250). He argues that a liberating education is one which abandons pedagogy (a philosophy for teaching children) and adopts andragogy (a philosophy for teaching adults) for learners of all ages. He argues that:

> ... unlike the case of pedagogy, the new techniques have to be premised both on the dynamic nature of society and that of the students and teachers—all of whom are in a constant process of maturation. ... Life is such an endless research problem that no student can ever come out of any educational institution with ready-made solutions to it. (Kabuga 1984)

This vision for liberating education is rooted in andragogy, a term coined by Malcolm Knowles, whose conception of andragogy is still predominant.

According to Knowles's seminal *The Adult Learner: A Neglected Species*, the principles of adult education are:

1. Adults are motivated to learn as they experience needs and interests that learning will satisfy; therefore, these are the appropriate starting points for organizing adult learning activities.
2. Adults' orientation to learning is life-centered; therefore, the appropriate units for organizing adult learning are life situations, not subjects.
3. Experience is the richest resource for adults' learning; therefore, the core methodology of adult education is the analysis of experience.
4. Adults have a deep need to be self-directing; therefore, the role of the teacher is to engage in a process of mutual inquiry with them rather than to transmit his or her knowledge to them and then evaluate their conformity to it.
5. Individual differences among people increase with age; therefore, adult education must make optimal provision for differences in style, time, place, and pace of learning (Knowles 1984).

From early on, andragogy was linked to distance learning, now used synonymously with online learning (Moore 2016). Both andragogy and distance/online learning regard the learner as being primarily responsible for managing their own learning, though for different reasons. For adult education, this stems from a conception of the learner as self-motivated, as having specific goals which education could help them achieve. For online learning, the distance of the learner also entails the learner's independence and greater self-reliance.

In this chapter, I will refer to adult education, and to pedagogy. I use "pedagogy" rather than "andragogy" because the primary form of education engaged with in this chapter is place-based education, commonly referred to as pedagogy of place. I also believe that the principles of adult education are appropriate for child and youth education as well, and that we are seeing those principles incorporated into mainstream public education pedagogy. Further, the term *pedagogy* is often a stand-in for a philosophy of teaching, no matter the age group.

In all cases, when I refer to education or learning, I mean *personal and professional growth that occurs through positive encounters with information, experiences, and other people*. When I refer to pedagogy, I mean the philosophies and techniques employed by the educator to help facilitate learning.

This chapter is written with the aim of encouraging and enabling readers to offer this type of education to others, and to demand this kind of education from your educators.

Planetary Health

> We cannot have an approach that is just about people or just about nature. People depend on nature but at the same time these activities that people realize are impacting natural resources. They need to understand that their activities are having an impact on these resources, but it's hard, because if they don't have another alternative, how can they live? How can they survive? It's challenging but I think we can do it. (Participant 2020-06 2020)

> I'm a really environmentalist person. I like to be an activist for nature and I really like the focus in poverty, so I made that connection. We live in a really mega diverse environment in Latin America, so climate change is really affecting our economies because farming and all our economic activities depend on nature. So, anything that happens with nature may affect our way of living. (Participant 2020-02 2020)

Planetary Health is a relatively new term whose definition comes from *The Rockefeller Foundation—Lancet Commission on planetary health* and has since been codified in the open-access journal, *The Lancet Planetary Health*. The Commission defines planetary health as "the health of human civilization and the state of the natural systems on which it depends" (Whitmee et al. 2015). At the Dahdaleh Institute for Global Health Research, where the Planetary Health Film Lab took place, it is described this way: "Planetary Health research calls urgent attention to the human health cost of environmental degradation and invites deeper reflection on the relationship between human and environmental wellbeing.... [It is] an urgent interrogation into the relationship between human civilization and the biosphere" (DIGHR 2018).

I would adjust these definitions. The primacy of human health betrays the anthropocentrism of this view of the planet. Referring to human society as "civilization" betrays the primacy of the sciences at the foundations of this concept. To my ears—shaped by the humanities and social sciences—"human civilization" is a scientist's way of trying to talk about people, akin to the animal kingdom. I am inclined toward more humanistic (though not anthropomorphic) ways of trying to talk about the planet.

Ultimately, these definitions of planetary health reinforce divisions between the human and everything else. This oppositional relationship is the result of highlighting humans' outsized impact on the health of the entire planet. We do have to face this, but I want to move away from any view which considers humans in the one hand, and the rest of the planet in the other. When I write of planetary health, and help shape an adult education program about it, I am working from a relationship-based conception. Planetary health refers to *the well-being of the whole of the planet, the complex web of relationships which shape that well-being, and the ways in which humans are embedded in those relationships.*

Pedagogy of Place

> I had a lot of opportunity to travel as a child and I noticed that what I saw as a child was no longer the same. Like after the oil spill in the Gulf of Mexico the biodiversity was not the same; it was completely different. And I just hold those memories very close to me.
>
> So, when I saw that, I thought, "that's not right," and when I started to travel to other places and notice that their ecosystems were not quite right [. . .] I started doing research on what is causing the ecosystems to shift so quickly. We know climate change is a problem, we know plastic in the ocean is a problem, but what I didn't realize until I started school is that the fashion industry is the second biggest polluter after oil. And it's a huge reason for ocean acidification, and temperature rise, and habitat contamination [. . .]
>
> Education and art are a great way to educate people about it [. . .] Not just as a young person, but for everyone, we're noticing that some changes need to be made and they need to be made now. UN and other policy-makers have the power to do that. And we know they've heard us, but the sooner they can start to make changes at a faster pace, it will really benefit not only the environment, but also the relationship that we all have.
>
> You know, people are saying that climate change is a young person problem, or an old person problem. I think it's everyone's problem, so I think if we all work together, like, I'm not a scientist, but if we get everyone to work together, we can come to a positive conclusion a lot quicker than we are. (Participant 2020-07 2020)

As with planetary health, place-based education is about relationships. Unlike planetary health, whose genesis formalized and canonized a specific definition, "place-based education lacks a specific theoretical tradition" (Gruenewald 2003). Rather, it draws from a number of pedagogical practices, theories, and thinkers, and can be described in terms of norms, trends, challenges, and principles. Broadly, it is a mode of teaching which centers on a physical location regularly accessible to the learner. In place-based education, encounters between a person and their place is indispensable (Smith 2002).

Pedagogy of place typically emphasizes ecological education, most often in rural settings. Another project I worked on, however, showed that pedagogy of place is relevant in every setting. In 2014–2016, *All I's On Education: Imagination, Integration, Innovation* asked, "What enables innovative pedagogies?" For eighteen months, we worked in ten Ontario schools that represented a broad range of the province's public education landscape:

rural, urban, suburban, isolated, secular, Catholic, anglophone, francophone, elementary school, middle school, and high school. A major finding was that a "sense of place (the historical, contemporary, and imagined context of the school and community) fosters culturally responsive and relevant inquiry" (Lundy and Kornberg 2016).

In a pedagogy of place, place is where, how, and why learning happens. Place is also what learners learn about. From this local starting point, the learner gets an education in the concepts and content knowledge fundamental to a global consciousness. Key to this pedagogy are positive experiences with place which foster positive, though often complex, relationships with place.

> Yeah, again, it's home [. . .] I've been thinking of it lately [. . .] It's a mining town. So, there's a lot of ideas about the wealth and the greed and something that our family is not really into. My parents are school teachers [. . .] Living in a place that like, not in a city, not many people, it opens your eyes, and teaches you to appreciate life.
>
> I mean, a lot people view [the town] differently. I've got friends who hate the town. They don't like it [. . .] The best thing to do tourist-wise is going to the shopping center or something like that. [My family's] favorite thing to do is kite surfing or go paddle boarding or paddling. That's our way of life. That's our fun. But the mining is out west, so we don't actually see the big effects of mining [. . .], but everyone knows it's there. Our mascot is a piece of coal.
>
> Yeah, we are the mining town. That's the things that you see. But the town in general is pretty spectacular. It's small, but when you see it from my aspect of it, it's a lucky place, we're just so lucky [. . .] I live in the water [. . .] And yeah, being in the ocean every day, you see plastic, you see animals washed up on the beach, you see things like that. You experience the heat and the change in temperature. A lot of people in life they just kind of forget it. Like it's just another hot day. (Participant 2020-04 2020)

PLANETARY HEALTH FILM LAB

> Yeah, I'm actually not 100 percent satisfied, I don't think any of us are, because we always want to be more accurate, more specific, to perfect everything, so we are not so content with everything. But we really love the process. Not only in filmmaking but in environmental issues, humanitarian issues, so we are proud of that, of the changes that we made in our lives coming here. It was just one week long, but we learned a lot. (Participant 2020-02 2020)

Film was once one of the more difficult media to work with. It required heavy, specialized, and expensive equipment, specialized knowledge, and usually multiple people. Once the film was made, it would have been difficult for anyone outside of the makers' immediate circle to watch it. New technologies have made films easier to make, new platforms have made them easier to share, and audience responsiveness to video has been high. Film has proliferated as a medium of storytelling and information sharing.

The Planetary Health Film Lab was conceived as a training ground for this generation's young climate advocates working in film. Among the seven participants, most had some experience with filmmaking or with environmentalism. Few had a lot of either; none had a lot of both. They all brought footage with them. The amount and type varied, but generally they provided previously shot B-roll (secondary footage relating to the content discussed by the interview subjects) and A-roll interviews. Over the course of five days, they attended learning sessions; they wrote their scripts; many did more filming, mainly interviewing researchers based at York University; they edited the films; they uploaded their films to the Youth Climate Report GIS map; and they presented their films at a screening event and reception.

This was no small feat. Anyone who has ever grappled with editing software knows that this alone could eat up all five days.

How We Got It Done

While this chapter's focus is on core principles of place-based planetary health education in a digital space, it is worth noting a few concrete elements that positively shaped this program.

Supportive Environment

> There were like five or ten minutes I was stressing out, because I have anxiety, and that was the moment when you and Mark and everyone was like, "It's okay, we'll sort it." There was no room for frustration, it's so chilled out here. (Participant 2020-05 2020)

In their exit interviews, a number of participants said that one of their main learnings was dealing with stress effectively, and that the support of program facilitators and of other participants was key. As educators, we can sometimes think there is a trade-off between process and product. The Film Lab's experience spoke to an approach which balanced the two by providing a lot of support. We stayed late in order to allow participants to keep working in the space. We made sure they took breaks. We joked, we encouraged, we gave feedback that focused on both the strengths and weaknesses of the work in progress, we brainstormed. We were on-hand, but we were not hands-on. We did not look over anyone's shoulder. We sought to understand each participant's vision for their work, asked them sometimes to consider other perspectives, and ultimately aimed to support their vision.

Semi-Structured Schedule

The workshop agenda was full. In addition to time to script, film, and edit the films, it included seven seminars; an evening in downtown Toronto; time for breakfast, lunch, dinner, and breaks; and a film screening and reception on the last evening. Seminars were in the mornings, and time to work on the films were slotted for the afternoons. This structure ensured everyone came together at the start of day and allowed the flexibility of self-managing in the afternoon, and often into the evening.

Attention to Logistics and Communications

This was a complex project logistically. Bringing together youth from all over the world meant facing time zone differences, visa applications, and wire transfers for reimbursements. That the participants were youth—many still students—meant that we could not in good faith ask them to pay for their flights and wait up to three months for their reimbursement to be processed. The Financial Officer at the Dahdaleh Institute took on the complex task of coordinating with each participant to book their flight on their behalf. We also covered the costs of visas, something we considered essential to equity.

Some of these challenges were not met. After visa issues, passport issues, and scheduling issues, seven of the twelve invited participants were able to come. A particular case of a would-be participant from Malawi was telling. In order to apply for a visa, he would have had to travel to South Africa to give his biometric information. We agreed to cover the associated costs, but there simply was not enough time. Make note: the visa process for Africans is an obstacle course, and sufficient time and funds should be set aside to ensure participation. We had participants from North America, South America, Europe, Asia, and Oceania. The only inhabited continent from which we had no participants was Africa.

During this process, program leads communicated with the participants. We had a Facebook group and we emailed. We asked them to provide more details about the stories they wanted to tell in their film. We were transparent with them as the program agenda developed, though this sometimes caused confusion and uncertainty. We provided a preparation guide with a program summary, travel arrangements, contact information, and a list of appropriate clothing for the Canadian winter.

These issues are worth noting, for much of the integrity of a program is in how much organizers dedicate to the boring stuff. Trust is built through the quality of the logistics, and the consistency of communications. Digital technologies, which would be offered as solutions to some of these challenges, come with their own logistical nightmares. If not addressed, they can make educational programs inaccessible and exclusive.

THE PLACE OF GIS WHEN PLACE IS ELSEWHERE

> It's one of those areas where anyone can watch [. . .] Film puts it all there. And with documentary film you can bring in writing and photography to help tell the story. Film is so available these days, everyone has a phone. You can pull your phone out and you can film anything and stich it all together for free and make an awesome video. That's how I started. It's available and it's right there and anyone can do it. (Participant 2020-04 2020)

The Planetary Health Film Lab was structured largely around two technologies. First, filmmaking and editing technologies. Second, the GIS map of the Youth Climate Report, hosted on Google Maps. This map hosts short documentary films created by youth around the world. Films typically feature a climate scientist who is interviewed by the young filmmaker, and the documentary serves as a science communication tool, aimed primarily at United Nations policy-makers.

Each film included on the Youth Climate Report is geo-located, and input with meta-data: filmmaker, title/category, x-coordinates, y-coordinates, URL to the climate researcher profiled, frame code of the film (which is uploaded to YouTube), country, and year.

The Planetary Health Film Lab's concrete outcome was seven original short documentaries that would be uploaded to the Youth Climate Report's GIS map, and the program was structured to facilitate that outcome.

There is a profound contradiction within an educational experience about planetary health that is at once centered on a digital space and that is also place-based. It is this contradiction that I will explore in the rest of this section.

On the one hand, a GIS map is one of the digital technologies best suited for a pedagogy of place. It allows learners to articulate the particularities of their own locales while putting it in relationship to other contexts globally. GIS mapping may be one of the best educational technologies to overcome the "tyranny of distant places" (Greenwood and Hougham 2015) and foster a truly global environmental consciousness. It marks participants on the "digital trail" (Greenwood and Hougham 2015) documenting the changing environment.

Given how GIS is at once a ubiquitous and largely invisible structure underlying many digital interactions and networks, it was also an enormous benefit to Planetary Health Film Lab participants to have learned about it, to be conscious actors when interacting with at least this one example of a GIS database. It is an aspect of digital literacy.

In all of these ways GIS mapping has the potential to be part of a rich place-based planetary health educational experience. There are no guarantees, however, that this potential will be met. GIS maps can play a role in building a global environmental consciousness, but they do not do so in themselves. Digital literacy is not the same as environmental or health literacy. And while GIS maps can be a space where participants leave a mark on the digital trail, leaving that mark does not in itself constitute meaningful learning. It is worth noting also that there is an ecological cost to these technologies (Greenwood and Hougham), and of the flights which brought participants to and from Toronto. There are benefits to these as well, which need to be weighed and evaluated.

Not only does the GIS map not guarantee a pedagogy of place; there are ways that the GIS map stands in direct tension with the conventions of place-based education. Where place-based education seeks "to create experiences where people can build relationships of care for places close to home" (Gruenewald 2003), the GIS map seeks to put science and stories in a global context. The relationship is enlarged, but the learner is sidelined. No longer is the learner the center of the educational experience. They are the interlocutor, a facilitator for the science-story. That science-story (the film), and its relationship to other science-stories from all over the world (the GIS map), are at the center of the experience. The focus on a technology can undermine the focus on education and pedagogy.

This GIS map in particular, the *Youth Climate Report*, seeks to emphasize what is similar across the all the films to convey a coherent and ultimately singular narrative about climate change. While this makes for a strong advocacy approach, it makes for a weak educational approach. An educational program which aimed at producing different versions of the same narrative lost opportunities to foster real growth and substantive learning. One film, for example, told the story of a pristine and beautiful landscape under threat by climate change. The truth is more complicated, and only came out in the exit interview, when the filmmaker explained that they came from a mining town whose mascot was a piece of coal. This filmmaker's place encompasses the beautiful rural landscape depicted in the film, and also the industrial landscape of the mines, and the suburban landscape of shopping malls; it encompasses class divides and complex social relations. None of this came through in the film. Even more problematically, none of it came through in the classroom.

Greatest of all challenges is that the GIS map stood in for the shared place that is at the heart of place-based education. Typically, this shared place is the physical location through which and about which learning happens. In the

Planetary Health Film Lab, the physical presence of those places was limited to moving images and sounds, saved as bytes, interacted with on computers, and completely unique to each participant. The shared physical space was the boardroom we used as a classroom, workroom, and breakroom. The boardroom, however, was not the place of our pedagogy. Instead, the digital space of the GIS map acted as a poor replacement for the in real life (IRL) experience of common ground.

Ultimately, the contradiction comes not from the involvement of technology, but from the disconnect of a place-based education program that occurs away from place. This is particularly problematic for a program about planetary health, not only because of its focus on the relationship between people and the planet, but also because it is about health. Health and illness are intensely personal experiences and corporeal. Our notions of well-being are as intrinsic to our selves as are our bodies (Martusewicz 2019). The great potential of a pedagogy of place for planetary health education is to tap into learners' sense of wellness and put it in relation to the well-being of the wider world. This cannot be done through a disembodied technology such as a GIS map, and it is difficult for educators to help learners build these relationships when educators and learners are not in place together.

Technology is not the problem, but it confuses the problem when it is offered as the solution, rather than as a tool to support creative approaches.

Components of a Planetary Health Pedagogy of Place, Anywhere

> The program is a tool and an important, pertinent space, where young people can tell their stories, as I mentioned, the youth who are tired of so much unconsciousness, so much lack of concern about these environmental issues, and know that these spaces are important and necessary to give consciousness to others and to create a real change. (Participant 2020-01 2020)

These and related challenges would be faced by anyone seeking to engage a pedagogy of place, mediated by digital technologies or constrained by funds, time, and other resources that prohibit experiencing place together. Often, technology would be offered as the solution to these latter challenges. I have sought to show that technologies are not in themselves the solution. They are, however, part of the reality of contemporary education, and can be leveraged toward rich educational experiences grounded in place.

I offer the following components as core to planetary health place-based education, and some ideas about how technologies, particularly GIS maps, can be supportive tools.

Content Knowledge

> All the while I knew that climate change was affecting me and affecting my mental health, but I didn't realize it was climate change, you know? I have more clarity of thought right now [. . .] I didn't have the clarity of thought that climate change is linked with all of this, so when I was asked to make a film, I realized that there is a link between climate change and mental health. I connected the dots. (Participant 2020-05 2020)

One of the reasons pedagogy of place is perceived as challenging, if not impossible, is because it is difficult to ensure learning of any particular content knowledge. It is both learner-centered and place-dependent. An educator cannot dictate a learner be interested in any particular lesson, nor can an educator guarantee that any particular lesson will be found in any particular spot in the woods.

This has led to an unfortunate perception that place-based education is soft on content learning. I argue, however, that content learning is absolutely embedded in a successful pedagogy of place. Content knowledge provides the concepts and information necessary to make sense of place; enriches the learner's understanding of place; and opens doors to questions. The learner's academic, economic, or other advancement is also largely dependent on being brought into this knowledge. These goals should not be discounted or derided; they are part of the purpose of education.

Pedagogy of place is not light on content learning, but it is risky. The educator doesn't know exactly what there is to learn, or where the learning will go. It is risky because the educator is also a learner. Place is transdisciplinary, unbounded by academic or professional categories. It challenges educators to draw on content knowledge in order to understand the world, rather than to narrow and dissect the world so that it can act as an example for a single discipline.

These risks are also the great possibilities of content learning in place-based education. Each experience is unique, each learning is new, each encounter may offer an unknown unknown. Content learning is no longer limited to preset lesson plans. Rather, content knowledge is everywhere, could be anything.

In this sense, the unparalleled access to information provided by the internet can be a great support. Didactic teaching can be largely replaced by learners' research. The educator can focus on teaching the critical thinking and media literacy skills needed for learners to conduct their own research effectively.

A carefully planned GIS map where learners can add their insights as part of the learning process can also be an aid. Categories and fields need to be broad enough to allow for all kinds of possible learnings, while specific enough to be useful. Attention to the information architecture and nomenclature of the map's underlying database is crucial for the GIS map to facilitate, rather than limit, learning.

Love

> I have always loved nature, since I was a kid. I used to go to a park near my home as a kid, and I miss being near all this nature, these green colours that make you feel better from the daily stuff that you do. I was part of scouts too, and to have this connection between nature and helping people, that could be the reason [. . .]
> It's kind of difficult for me to communicate with people and say "Hey! Take care of nature!" Because for me, it's just a feeling that I love, and I don't need more reason. Nature has its own value for me, but I'm aware that for other people maybe it's not like that, maybe because they have different experiences. So, I want to tell them in a different way about the services that nature provides us. We have to look for the way that we can convince people, persuade them. I wish people just felt like taking care of [nature]. But there are other important aspects, so have to try that a different way. (Participant 2020-06 2020)

Pedagogy of place has long embraced positive relationships to the learner's locale as the basis for transformational learning (Gruenewald 2003). It also emphasizes action, and the learner taking on the work of bettering the ecology of their environment (Smith 2002).

I would call this love and claim that a pedagogy of place can be enacted in any learning context where a love ethic presides. As bell hooks explains,

> Embracing a love ethic means that we utilize all the dimensions of love—"care, commitment, trust, responsibility, respect, and knowledge"—in our everyday lives. We can successfully do this only by cultivating awareness. Being aware enables us to critically examine our actions to see what is needed so that we can give care, be responsible, show respect, and indicate a willingness to learn. (hooks 2000)

Hooks tell us that love is not a feeling, but a verb (hooks 2000). To love is an act of self-growth for the sake of nurturing another's growth. Education, as I have defined it, is a process of growth through positive encounters with information, experiences, and other people. Education is a matter of cultivating the awareness hooks writes about, which enables us to act lovingly. Love also speaks to a slow pedagogy, "which emphasizes a dwelling in place over time, as opposed to fast-paced, tiered demonstrations of skill within wilderness environments" (Bertling 2019).

In planetary health education, I argue, a love ethic means that the educator's role is to support the learner as their relationship to place matures, and as they seek to actively and positively participate in their place.

A pedagogy of place which fosters a loving relationship between the learner and their place must primarily happen in that place. Digital technologies can help. Bradley Garrett, author of *Explore Everything: Place-Hacking the City*, has described how urban explorers use cameras to connect with the places they walk through (Garrett 2012). Though cameras are sometimes perceived as a shield or barrier between the photographer/filmmaker and the world, Garrett (2012) claims that the camera is often the reason to slow down, to look closer, to consider more carefully. Filmmaking can be an exercise in noticing, as well as a way to trace a learner's journey from first encounter to greater understanding.

A film from the Planetary Health Film Lab exemplifies this. The filmmaker's journey started with seeing news footage of flooding in his home city of Venice. Filmmaking was his way of asking why, his reason to read and watch more news reports, his reason to talk to experts, his reason to get on a boat and look at the gates around the lagoon. He worked through the complexities of municipal politics and global climate change and came to his own understanding of why the floods happened. He did this because he was making a film about it, and needed to make his contribution to the Youth Climate Report.

> That was the biggest learning, that even as a filmmaker you can do something about climate change, and you can use your tools to tell your stories while doing good. (Participant 2020-03 2020)

Storytelling

> Working in the team. You have Mark [. . .] and all the other guest speakers who've had a lot of experiences, and also the participants, from all over the world, with the different perspectives and the different stories to tell. It all helps to build up your own story, to keep ideas flowing. (Participant 2020-04 2020)
>
> A lot. I learned about the other participants, about their cultures, their problems, and I saw how this united all of us, that we can generate a change, something big, something important, so that everyone can know what we are facing. So, yes, there was much learning, much understanding. (Participant 2020-01 2020)

A love ethic compels place-based educators to ensure learners spend time in their place. When learners and educators come together in a shared space—be it physical or virtual—they need to bring their experiences back. They need to tell stories of place to each other.

As we saw with content knowledge, place is transdisciplinary. Stories of place are not bound by academic subject siloes, but can encompass science, history, sociology, ecology, and so much more. Stories can be a productively messy blend of different ways of knowing, can draw equally from family history and botanical sciences to describe a place more fully (Kimmerer 2020).

As part of the process of storytelling, learners can read, listen, and watch stories that have been previously added to a GIS map like the Youth Climate Report. The learners' completed stories—in this case, as short films—can be added to that GIS map. Telling stories of place on a larger platform is a form of advocacy, of claiming and contesting narrative, of advocating for action (Somerville 2010).

But it is the intimate storytelling which takes place between learners and educators which is crucial to the personal and professional growth that defines education. Storytelling is so important because it consists of a fuller, connected, fluid, complex self-conception that is not divided into discreet points of data, as a GIS database is.

Storytelling is an act of place-making. It is through telling stories of place to others in an educational experience that a learner makes sense of their relationship to place. It is through telling stories of place to a wider audience, on an elevated platform, that a learner acts to shape the wider narrative of that place.

> I think that having to work with people with different perspectives with different kind of side of the story and seeing how they can manage to solve any problem from where they are. That's an interesting thing to know, that wherever you are in the world you can make change, you can do something to make people conscious about a problem, or you can even solve a problem, because we are all part of a system. So just moving one piece of the system can make a big change [. . .]
>
> I would like to be a storyteller. I think it's really important for everyone to have something to fight for and to try to convince others that what you're fighting for is worth fighting for. That's why you need to be a storyteller [. . .] to know how to manage your ideas into something that could be powerful for everyone. (Participant 2020-02 2020)

CONCLUSION

In this chapter, we have tried to show that a pedagogy of place is the right fit for planetary health education. Place is transdisciplinary, as is planetary health as a field of study. Place-based education is both corporeal and conceptual, as is planetary health. Place-based education is about building loving relationships with place; planetary health is strengthened through such loving relationships.

This chapter is also about grappling with the role of GIS in contemporary place-based education for planetary health.

When GIS technologies take the role of place in place-based education, they undermine the embodiment of a place-based educational experience, and the capacity of the educator to foster loving relationships to place. These are natural tensions between the networked, data-based, digital space of the global GIS map and the fluid, story-based, intimate place of learning.

The tensions between GIS technologies and place-based education are fundamental, but not damning. GIS maps offer a platform, "an architecture from which to speak or act" (Gillespie 2017).

> Figuratively, a platform is flat, open, sturdy. In its connotations, a platform offers the opportunity to act, connect, or speak in ways that are powerful and effective: catching the train, drilling for oil, proclaiming one's beliefs. And a platform lifts that person above everything else, gives them a vantage point from which to act powerfully, a raised place to stand. (Gillespie 2017)

A platform, however, is a tricky metaphor for digitized networks of information databases. It gives the impression or feeling of power, of connection, of elevation, but it is unclear to what extent anyone is actually powerful, connected, and elevated by using any particular platform. Indeed, many networked digital platforms, such as social media, are far from "flat, open, sturdy."

Nonetheless, the platform metaphor is useful for the ways that it highlights how a GIS map is a public viewing room where learners can offer what they have learned to others. It is a space in which the learner's work can live, linked to the learning work of others. A pedagogy of place for planetary health education is absolutely possible with, and can be bolstered by, GIS and related digital technologies.

To close, I want to revisit a quote from one of the participants we saw earlier. The participant's insight into her own motivation for participating in the Film Lab, and why she cares about climate change, brings together all the components we have explored: place, content knowledge, storytelling, advocacy, learning, experience, relationships, and, above all, love.

> I have always loved nature, since I was a kid. I used to go to a park near my home as a kid, and I miss being near all this nature, these green colours that make you feel better from the daily stuff that you do. I was part of scouts too, and to have this connection between nature and helping people, that could be the reason [. . .]

It's kind of difficult for me to communicate with people and say "Hey! Take care of nature!" Because for me, it's just a feeling that I love, and I don't need more reason. Nature has its own value for me, but I'm aware that for other people maybe it's not like that, maybe because they have different experiences. So, I want to tell them in a different way about the services that nature provides us. We have to look for the way that we can convince people, persuade them. I wish people just felt like taking care of [nature]. But there are other important aspects, so have to try that a different way. (Participant 2020-06 2020)

BIBLIOGRAPHY

Bertling, Joy G. "Non-Place and the Future of Place-Based Education." *Environmental Education Research* 24, no. 11 (2019): 1627–30. https://doi.org/10.1080/13504622.2018.1558439

DIGHR. "Planetary Health." Dahdaleh Institute for Global Health Research, 2018. https://dighr.yorku.ca/research/planetary-health.

Garrett, Bradley L. "Place Hacking." Seminar Series, Centre for Research in the Arts, Social Sciences and Humanities, Cambridge University, October 29, 2012.

Ghebreyesus, Tedros Adhanom. "WHO Director-General's Opening Remarks at the Media Briefing on COVID-19." World Health Organization, Geneva, March 11, 2020. https://www.who.int/dg/speeches/detail/who-director-general-s-opening-remarks-at-the-media-briefing-on-covid-19.

Gillespie, Tarleton. 2017. "The Platform Metaphor, Revisited." *Digital Society Blog* (blog), Alexander von Humboldt Institute for Internet and Society, 2017. https://www.hiig.de/en/the-platform-metaphor-revisited/amp/.

Greenwood, David A., and R. Justin Hougham. "Mitigation and Adaptation: Critical Perspectives toward Digital Technologies in Place-Conscious Environmental Education." *Policy Futures in Education* 13, no. 1 (2015): 97–116. https://doi.org/10.1177/1478210314566732

Gruenewald, David A. "The Best of Both Worlds: A Critical Pedagogy of Place." *Educational Researcher* 32, no. 4 (2003): 3–12. https://doi.org/10.3102/0013189X032004003

Hirsch, Brett D. "Digital Humanities and the Place of Pedagogy." In *Digital Humanities Pedagogy: Practices, Principles and Politics*, edited by Brett D. Hirsch. Cambridge, UK: Open Book Publishers, 2012.

hooks, bell. *All About Love: New Visions*. New York: Harper Perennial, 2000.

Kabuga, Charles. "Andragogy in Developing Countries." In *The Adult Learner: A Neglected Species*, edited by Malcolm Knowles, 250–56. Houston: Gulf Publishing Company, 1984.

Kimmerer, Robin Wall. *Braiding Sweetgrass: Indigenous Wisdom, Scientific Knowledge and the Teachings of Plants*. Minneapolis: Milkweed Editions, 2020.

Knowles, Malcolm. *The Adult Learner: A Neglected Species*. 3rd ed. Houston: Gulf Publishing Company, 1984.

Lundy, Kathleen Gould, and Netta Kornberg. *Research Findings and Recommendations for All I's on Education: Imagination, Integration, Innovation.* Council of Ontario Directors of Education and the Ontario Ministry of Education (Toronto), 2016.

Martusewicz, Rebecca A. *A Pedagogy of Responsibility: Wendell Berry for EcoJustice Education.* New York: Routledge, 2019.

McGregor, Deborah. "Decolonizing Global Health Research." Seminar Series, Dahdaleh Institute for Global Health Research, Toronto, Canada, September 30, 2020. https://dighr.yorku.ca/event/decolonizing-global-health-research-i-dr-deborah-mcgregor/?instance_id=101.

Moore, Michael Grahame. "Flipped Classrooms, Study Centers Andragogy and Independent Learning." *American Journal of Distance Education* 30, no. 2 (2016): 65–67. https://doi.org/10.1080/08923647.2016.1168637

Pavlovskaya, Marianna. "Critical GIS as a Tool for Social Transformation." *The Canadian Geographer* 62, no. 1 (2018): 40–54.

Smith, Gregory A. "Place-Based Education: Learning to Be Where We Are." *Phi Delta Kappan* 83, no. 8 (2002): 584–94. https://doi.org/10.1177/003172170208300806

Somerville, Margaret J. "A Place Pedagogy for 'Global Contemporaneity.'" *Educational Philosophy and Theory* 42, no. 3 (2010): 326–44. https://doi.org/10.1111/j.1469-5812.2008.00423.x

Whitmee, Sarah, Andy Haines, Chris Beyrer, Frederick Boltz, Anthony Capon, G., Braulio Ferreira, Souza Dias, and et al."Safeguarding Human Health in the Anthropocene Epoch: Report of The Rockefeller Foundation-Lancet Commission on Planetary Health." *The Lancet* 386 (2015): 1973–2028. https://doi.org/10.1016/S0140-6736(15)60901-1

Chapter Ten

Geomedia in the Classroom
A Pedagogical Approach to GIS-Enhanced Ecocriticism

Mark Terry, Erik Tate, and Shahreen Shehwar

Geomedia is ubiquitous. We find it whenever we order an Uber or a pizza; take a picture with our smartphones; post on social media; or track our flight progress as we sit on a plane. We can drive to any destination without needing directions and find the closest Mexican restaurant to wherever we might be at the moment.

With all this geo-locative technology in our lives, it becomes apparent that we need to incorporate it into our educational programs to understand how to best use it as well as to use it as a tool for teaching and learning. Global issues related to the environment seem like the logical place to start, with analysis of time and space yielding new data (Terry 2020) being a unique affordance of geographic information systems (GIS) as well as providing a panoptic perspective of existing data. The environmental humanities and ecocritical study can benefit from a pedagogy specifically related to the increased availability of free-to-use or open source and internet-based GIS mapping resources to achieve this goal.

While QGIS is a free GIS software platform with functionality equivalent to ArcGIS proprietary software, KnightLab's StoryMap JS, GoogleMaps, Harvard's WorldMap, OpenStreetMaps, CartoDB, Mapbox, and many others offer instructors a variety of user-friendly, customizable, free or price-tiered, proprietary or open source, tools for creating maps and utilizing geospatial data in social research. While traditional GIS programs, such as ArcGIS and QGIS, can be difficult to master, newer forms of GIS make mapmaking and data management more accessible to nonspecialists and are easier to learn.

In the last decade, GIS and digital mapping software has become decentralized, simplified, and distributed among a vast array of online platforms, map databases, and geotechnologies known as the Geoweb, whereas before GIS was largely dominated by resource-hungry and technically demanding proprietary software suites (Pavlovskaya 2018). These developments point to new pedagogical space beyond a singular emphasis on developing technical skills and training GIScience practitioners. They make room for teachers to introduce students to a critical approach that centers on instilling an awareness of 1) the history and social consequences of GIS and related geospatial technologies, 2) the broader critical-humanist discourse around the social production of spatial knowledges, and 3) the potential to transform social relations and remake spaces by using GIS in and beyond the classroom. Likewise, critical GIS pedagogy means not only teaching students how to use GIS, but also to reflect on what GIS is, its broader social implications, and the students' own positionality and situatedness in relation to these technologies and applications.

The social, political, and economic challenges posed by the global ecological crisis, and the increasing visibility and salience of environmental issues in the news media, intergovernmental conventions, academic conferences—not to mention literature, film, and television—has created a unique opportunity for ecocriticism to emerge as a critical paradigm shaping the theory and use of GIS in the classroom. The use of GIS already has a foothold in the humanities outside geography (for instance, as spatial humanities GIS, digital humanities, and historical GIS), but its role in ecocinema and ecocritical pedagogy remains to be explored.

Nevertheless, it is important to acknowledge that the convergence of GIS and environmentalism has already begun outside of the university. ESRI and National Geographic have partnered to promote a yearly StoryMaps competition to promote its environmental application among secondary school students (ESRI, 2021). Likewise, the Youth Climate Report (YCR), with support from the United Nations Framework Convention on Climate Change (UNFCCC) and the United Nations Educational and Scientific Organization (UNESCO), hosts several annual workshops, training sessions, and competitions on various topics worldwide, the winners of which are added to the YCR geographical database documentary (Geo-Doc) platform. All workshops and competitions ask students and young filmmakers aged eighteen to thirty to engage with the United Nations' Sustainable Development Goals (SDGs) and to use data-driven storytelling and documentary film-based approaches to climate change issues. Their success at giving youth a voice in climate discourse, a chance to exercise creative and collaborative storytelling, and a platform for their community activism and research is testament to the future potentialities of GIS in postsecondary classroom education. From an ecocritical perspective, GIS in general, and the Geo-Doc as a particular GIS-enhanced mode of documentary filmmaking, can enhance learning and project-based engagement, is versatile and user-friendly, inspires critical thinking, and for teachers, and facilitates the instruction of global environmental issues.

HISTORY AND CONTEXT

The usage of geographic information system (GIS) technologies in classrooms is by no means a new frontier in many places outside of Canada. The Danish in 1999 were perhaps the earliest pioneers of GIS programs, with ArcView 3.0 becoming standard for geography teachers at that time (Jensen 2011). In Denmark, many geography-oriented IT programs had been developed since the early 1990s, many of which were "conceived and developed by individual enthusiasts" (ibid). By contrast, in Canada, while classrooms now have free access to GIS technologies, the full potential of classroom GIS is only just beginning to be explored in many provinces and territories.

More recently, the province of Manitoba in Canada unveiled its Better Education Starts Today (BEST) strategy for the K–12 curriculum, which includes a provincial site license for ESRI Canada GIS software, ArcView 10 (GoM 2021). This program was developed in response to the teaching challenges that came with the COVID-19 pandemic that pulled students from a standard classroom setting into a home-based, virtual teaching environment. The goal of this program was to develop an accessible K–12 curriculum, where students could succeed despite where they lived. Manitoba's choice to lean toward GIS learning at a time where learning has become more virtual speaks to the potential of classroom GIS to teach critical skills such as problem-solving, critical thinking, data literacy, and collaborative learning through programs that are universally accessible and relatively easy to learn. Furthermore, while classroom GIS in Canada is typically only connected to the discipline of geography, increasingly classroom GIS is being seen as being applicable to other subject areas in Canada, and around the world, particularly in the social sciences.

Why is a pedagogy for GIS in the classroom important? Because by employing the use of geomedia as a teaching tool for instructors and as a presentation and study tool for students, there is evidence that suggests students improve their overall critical thinking skills. Joseph J. Kerski and Thomas R. Baker explain this research in their paper "Collecting Geo-Data to Support Classroom Field Studies":

> Students who are well grounded in the geographic perspective through geo-technologies are better able to use data at a variety of levels, in a variety

of contexts, think systematically and holistically, and use quantitative and qualitative approaches to solve problems. In short, as graduates, they are better decision-makers. (Kerski and Baker 2014)

This chapter explores how the Geo-Doc is being used in the classroom to teach (eco)critical skills regarding global environmental issues through a two-year study of students and teachers in a first-year undergraduate course known as EU/ENVS 1010: Introduction to Environmental Documentaries offered at York University in Toronto, Canada. It also explores how students who lack knowledge of either digital mapping platforms or specific environmental issues respond to their first exposure to them in the classroom. There is a significant gap in knowledge production for classroom GIS, particularly in the Canadian context. As it stands, both the "negative and positive outcomes of GIS use in the classroom are largely unclear" and "with little research evidence as backup" (Siegmund, Volz, and Schulman 2007). Many geography teachers, for instance, are not aware of how to use GIS, much less how it can function as a teaching tool (De Lange 2006, cited in Siegmund, Volz, and Schulman 2007). Furthermore, Canada presents a unique environment for studying the use of classroom GIS. Canada has a "decentralized curriculum landscape," where the "governance of education is a constitutional right given to the ten provinces and three territories," each guided by "its own curriculum developed and overseen by their respective Ministry of Education" (Huynh et al. 2017). With the expansion of classroom GIS in Canadian classrooms, it is also evident that there are substantial gaps in digital literacy among Canadians due to factors such as "Canada's lack of institutional response to digital literacy needs in education" and the "lack of a national strategy" (Hadziristic 2011). As such, studies like this can promote GIS more universally among Canadian provinces and territories as an essential in-class learning tool and also promote data literacy more universally in classroom learning environments.

This chapter includes discussion of a case study to unpack some of the overarching themes that were identified over the course of the study and remarks on some of the benefits in exposing both domestic and international students to classroom GIS and also explores classroom GIS as a qualitative and quantitative learning tool, going through how classroom GIS can effectively investigate contemporary issues at a time when understandings of "space" and "place" are shifting, and it is becoming increasingly necessary to have a spatialized understanding of socioecological issues.

CLASSROOM CASE STUDY

The *New Media Approaches for Environmental Studies* is a multiyear study that looked at how York University students responded to the Geo-Doc creation component of a course titled, EU/ENVS 1010: Introduction to Environmental Documentaries. The course introduces students to "how environmentalists are using documentary films for speaking truth to power" (Terry 2021). Covering a range of filming techniques and deconstructing the environmental documentary, students found the course interesting not only for those studying film, but also for those who are studying environmental issues, especially how these issues can catch public attention through presentations made in a digital multimedia project. The course was designed with an emphasis on balancing documentary theory, environmental activism and communication, and experiential project-based learning centered around the creation of a Geo-Doc that addresses one or more of the United Nations' Sustainability Development Goals.

The study was conducted to help in the assessment and improvement of the Geo-Doc as a pedagogical tool for use in the classroom. The study's results show how new users of classroom GIS benefit from the combined functionality of Google My Maps and YouTube as a pedagogical tool. The students performed tasks such as setting up CSV files as the information database for their projects using Microsoft Excel or an equivalent program, determining and formatting coordinates for each entry according to Google My Maps specifications, and properly embedding video links in map pins. Although most students had never used GIS or digital mapping software as a data management or visualization tool, knowledge of how to use professional mapping software did not present a significant advantage when introducing students to the Geo-Doc and Google Maps. This is because Google Maps, like many digital mapping platforms, is user-friendly, easily customizable, and does not require the same extensive training as desktop software–based GIS suites. The study revealed that this exposure to GIS, through creating a Geo-Doc, was a positive experience, with one student planning to use it for future project presentations in other courses and one teacher planning to use in presenting lessons on global issues. The study indicated that students understood the potential applications for hybrid, multimedia mapping projects in the environmental humanities and other fields such as economics, ecology, communications, history, education, and geography, with others likewise speculating that the Geo-Doc would be a useful tool in these areas.

Since the study began before the pandemic and continued afterward, there were two distinct learning environments within which both teacher and student groups were engaged. The first was the traditional classroom structure at the postsecondary level: the professor delivered a weekly two-hour lecture in a 200-seat lecture hall followed by classroom tutorials led by teaching assistants for twenty-five students each. When the pandemic descended, we were all under lockdown and the course was taught remotely. Lectures were delivered asynchronously while tutorials took place synchronously via Zoom, a digital communications platform with which we are all too familiar.

In year 1 of the study, there were 200 students; in year 2 during the pandemic, there were eighty. The authors of this chapter, as well as being researchers in the study, were also part of the teaching team of three for the courses. The professor did not participate in contributing to the data through questionnaires as he also served as the study's principal investigator; Mr. Tate and Ms. Shehwar provided their data based on their teaching experiences related to the study's questions. The questions posed to both teachers and students were as follows:

For Teachers:

1. Were you familiar with GIS software before this project? Which ones?
2. Have you ever made a GIS map before? For what purpose, and if not, why not?
3. What challenges did you face teaching GIS?
4. How would you use the Geo-Doc as a teaching tool in other courses?
5. Now that you know how to make a Geo-Doc, do you think you will use this technology again as a teaching tool for other courses? Please provide an example.

For Students:

1. Were you familiar with GIS software before this project? Which ones?
2. Had you ever made a GIS map before? If not, why not?
3. What challenges did you face creating your Geo-Doc project?
4. What did you learn in creating this project?
5. Now that you know how to make a Geo-Doc, do you think you will use this technology again for other projects in other courses? Please provide an example.

Key Findings: Teachers:

For question 1, 50 percent of the instructors interviewed were familiar with GIS software prior to teaching the ENVS 1010 course used in this study. Two respondents cited experience with ArcGIS.

For question 2, all instructors claimed to have made a GIS map despite only half of them stating previously that they were familiar with GIS software. The maps they claimed to have made were maps created for in-class instruction for previous iterations of ENVS 1010 which they had taught.

For question 3, the greatest challenge in teaching GIS came from trying to communicate with students remotely during the pandemic. While the GIS mapping platform (Google My Maps with integrated YouTube functionality) apparently presented no inherent pedagogical challenges to respondents, 50 percent of respondents pointed to the virtual learning environment as presenting difficulties. Ensuring students can perform the essential tasks of 1) setting up CSV files as the information database for their projects using Microsoft Excel or an equivalent program, 2) determining and properly formatting coordinates for each entry according to Google My Maps specifications, and 3) properly embedding video links in map pins, so that they will play directly on the map without redirecting the user to an external site, requires more time in the virtual classroom.

For question 4, all respondents said they would use GIS and Geo-Doc platforms as a teaching tool in other courses. Fifty percent of respondents said it would be best applied to social science courses; 25 percent said it would best applied in political sciences courses; while the remaining 25 percent believed it is best suited for other (unspecified) courses. Three-quarters of the respondents speculated that the Geo-Doc would be a useful tool to present to students in social science courses, and one respondent thought the Geo-Doc could be a useful tool for political campaigns in the context of electoral politics. One other respondent suggested the Geo-Doc has a range of useful applications.

For question 5, 50 percent of respondents said that they would use the Geo-Doc and GIS technology in future teaching assignments, while the other 50 percent were unsure. It is important to note that no respondent said they would not be using geomedia in future teaching assignments. Specifically, half of the respondents thought the Geo-Doc could be a useful presentation tool for presenting geospatial data or analyzing global issues.

Key Findings: Students:

For question 1, 70 percent of students interviewed claimed to have no experience with GIS technologies while 30 percent claimed they did. For the same question, 32.5 percent expressed awareness of consumer products utilizing GPS, location services, and web mapping. "Yes, I was aware of GIS software before this project like your smartphone GPS location apps (Google Maps) food delivery tracking, airplane tracking, etc."

For question 2, 92.5 percent of respondents said they had never made a GIS map before, while 5 percent said they had, and another 2.5 percent were not sure if they had or had not. The most common reasons given were: 1) They did not know how (perceived difficulty or lack of training); 2) They never had a reason to (not used in their program of study, no relevant assignments); and 3) They were unaware they could easily do so (unaware of existence of open source, custom mapping platforms).

For question 3, there were a wide range of responses expressing different areas of challenge in learning GIS reflecting different learning levels for students in general, as well as specifically between those students with previous GIS experience and those without. The areas identified by students as challenging were:

- Understanding the GIS software (Google My Maps): 23.4 percent
- Data collection: 23.4 percent
- Sourcing longitude and latitude coordinates: 19.1 percent
- Choosing topics or themes for the project: 17 percent
- Using Excel spreadsheets and saving as CSV files: 8.5 percent
- Data verification: 4.3 percent
- Working in groups with other students: 4.3 percent

Many students cited difficulties choosing a topic and framing a Geo-Doc or gathering and managing information or doing the research. Many students seemed to find the amount of information to be daunting. One students reported this challenge this way:

> It is really difficult for me to investigate various causes and cases of climate change in the process of producing the GIS project (because) I wanted to distribute the pins evenly on our map, so I looked for diverse sources, climate action videos made worldwide, but it was exacting and complicated because they were located in (many) specific areas. Therefore, the diversity and widespread distribution of metadata have been a great challenge for our team.

Another student responded similarly:

> When I created Geo-Doc, I needed to search a lot of information on the Internet that I had never seen before, and it allowed me to learn a lot about Geo-Doc. I think it will be a lot easier when I do it next time.

The responses indicate a lack of experience in general, with big data, and similar database projects and assignments in the classroom. However, despite this initial anxiety in dealing with large volumes of data, the students were able to complete their projects.

For question 4, the learning outcomes for students were divided into four areas:

- How to use and make GIS projects (45.7 percent)
- Global environmental issues and the United Nations' Sustainable Development Goals (28.3 percent)
- Film as a communications tool (13 percent)
- How to work in groups with other students (13 percent)

While nearly half of all respondents acknowledged learning how to use GIS as a classroom project platform, almost another third indicated they had learned a lot more about the subject of the project: environmental issues. Many indicated that their understanding of issues like climate change and the UN's Sustainable Developments Goals were enhanced through the global perspective provided by the GIS platform. Many others attributed their enhanced understanding of environmental issues to the film component of the project and the way the videos provided visible evidence of climate impacts around the world.

For question 5, more than half of all students interviewed (62.5 percent) said they would use GIS technologies and specifically the Geo-Doc platform to present projects in other courses, especially ones that are global in scope. While 25 percent were unsure, only 12.5 percent said they would not use it in other courses. Many said that they would be open to creating Geo-Docs in the future and indicated that it could be useful in specific courses (economics, geography, education, history, and ecology).

Generally, since the student participants mostly did not have previous exposure to digital mapping, the study's results were used to help improve learning outcomes for inexperienced map creators. For instance, many students indicated difficulties at the outset with choosing and conceptualizing an environmental research topic through the Geo-Doc platform, because many students lacked background knowledge of environmental issues. This difficulty was partly resolved by requiring students to frame their research around one or more of the United Nation's SDGs. Plenty of examples of specific environmental issues were also provided, so that students were more comfortable picking a topic and researching it.

Theoretically, the Geo-Doc was presented as a living documentary, meaning that it was not to be viewed as incomplete or complete, but as an ongoing, open-ended collaborative project. The final Geo-Doc that the students presented was not meant to be an exhaustive resource that covered everything on a topic, but rather something that was exploratory and could support ongoing research about a certain topic. In other cases, broader trends in higher education seemed to be behind certain difficulties with the Geo-Doc project; much of the students' inexperience with digital mapping is due to not encountering courses or assignments that required them to use digital mapping software. More specifically with Google My Maps, some students had difficulties inputting data and faced a significant first-time learning curve in using CSV (comma-separated values) functionality with the mapping platform. Students encountered difficulties in properly formatting longitude and latitude coordinates on a spreadsheet and embedding files directly on the map. That many students were not familiar with digital mapping applications is in line with the literature that finds classroom GIS, again, is only beginning to be explored in Canada.

Despite these difficulties, throughout the years that this study was conducted, students presented many exceptional Geo-Doc projects that viewed environmental issue from a multilens perspective, deconstructing the issue into several applicable subissues. For example, a student that presented on global hunger addressed gender disparities in hunger, technical solutions, as well as how climate change contributed to food insecurity around the world.

CLASSROOM GIS ENCOURAGES QUALITATIVE AND QUANTITATIVE LEARNING

In working together to create a Geo-Doc, students were exposed both to qualitative and quantitative understandings of GIS. They understood how the platform could be used to represent diverse geographies of knowledge—revealing the social realities lived, perceived, and experienced by different groups of people around the world. At the same time, the Geo-Doc taught students some core concepts of geographic information science, such as the creation of digital data, the use and access of existing data, and the combination of data (Albrecht 2007). They learned to compare analyze points on the Geo-Doc containing spatial data, while also plotting these nodes on the platform using their geographic location (i.e., longitude and latitude). One student, who chose the Sustainable Development Goal of Quality Education, stated:

> I believe that the GIS program can effectively express the unequal education of the world. It can discuss the differences in intercontinental or inter-regional education and the problems and causes that aggravate it. It means we can compare the reason why inequality of education is occurring. I also predict that it will be easy to explain the importance and efficiency of education.

The Geo-Doc helps students understand how different issues can be experienced differently around the world. This is highly useful, given that our understandings of "space" have shifted beyond "areal distinctions" toward more "socially situated threads of space as represented, perceived, and lived by different groups" (Battista and Manaugh 2018). Among the reasons that our understandings of space have shifted and become more analytically important is because of factors that have increased patterns of migration worldwide, such as globalization and climate change. Additionally, significant global issues, such as the Sustainable Development Goals, have made it more necessary to track social issues geographically, in order to gain a deeper understanding of how these issues are experienced around the world.

It is often debated whether online learning can be an effective retention tool for international students. On the one hand, remote learning excludes the cultural components of studying abroad. Many students are not as exposed to English conversations, for instance, and learning English may have been one of the reasons that these students chose to study in Canada. At the same time, the pandemic has made it necessary to revitalize the learning environment, which can include "course design based on machine learning to personalize student experiences" (Schrumm 2020).

The Geo-Doc's treatment of the humanities and social issues helps expose students to knowledge produced in other places. In the ENVS 1010 course, students shared many videos produced by individuals living in the Global South that presented primary understandings and lived experiences of the issues that they were studying. Many of these students did not enter this course with such nuanced understandings of place, but the Geo-Doc helped them to visualize complex social problems from the intersection of space and society.

At the same time, the Geo-Doc's ability to engage humanistic issues on an accessible and easy-to-use platform helps students develop a cursory understanding of the very complex spatial thought that GIS is capable of. It allows students an opportunity to engage with the possibilities and limitations of storing and acquiring spatial data, by compelling students to "gauge data reliability and evaluate the effectiveness of their analytical approach" but also "fostering creativity and quantitative research skills" that can be used in other applications (Battista and Manaugh 2017). As reflected in the surveys, this was an observation that the students had made themselves—recognizing that the Geo-Doc could be used in other courses and to cover other fields of study. Some participants saw the Geo-Doc as a valuable communications tool in their future careers in business to represent company interests that are global in nature.

Teachers involved in the study found a similar application for the Geo-Doc as a teaching tool. "For any international subject, geo-locating videos related to that subject provides a unique global perspective to the learning experience," said one teaching participant. "And by applying the analytical affordances of time and space, students will discover new data related to the issue thereby enhancing the learning experience through direct student engagement with the Geo-Doc platform."

If universities can explore opportunities to revitalize the virtual learning environment, language retention for international students can also potentially be improved in a virtual classroom. This is largely dependent on university offerings for international students. York University's English Language Institute (YUELI), for instance, offers extra learning opportunities, such as online social and cultural activities and undergraduate student partners, while also providing twenty classroom hours per week to help international students build a solid foundation for English proficiency. With the Canadian Bureau for International Education (2020) finding a 135 percent increase in international students in Canada between 2010–2020 (CBIE 2020), ensuring that international students are not left behind in the virtual classroom setting will be a foreseeable challenge for Canadian universities for the length of the COVID-19 pandemic. Virtual classrooms, if effective, can also part of a regular learning environment post-COVID, making it more important to ensure that they continue to be accessible and supportive for all students.

CONCLUSION

The Geo-Doc can be a very useful teaching and learning tool, even in a virtual classroom, and for international students. While the ENVS 1010 course is largely based on visual learning methods, the classroom GIS that was utilized in the course is easily adaptable for other courses and teaches critical learning skills, including data literacy, critical thinking and problem solving, and collaborative learning. It also presents an alternative way to gauge student learning from more traditional options such as quizzes or essays, which can be less engaging and, as such, less retentive. While both filmmaking and GIS software are more widely accessible today because of technological innovations and open source options that are available online, postsecondary institutions also need to do their part in promoting and shaping the use of these promising technologies and integrating them into course curricula beyond GIS specialization or certificate courses.

To encourage the use of classroom GIS in postsecondary institutions, more research should be done on the Canadian model, exploring the use of classroom GIS, including the Geo-Doc, in other provinces or territories to gauge what challenges and opportunities may arise. Post-COVID, when more in-class learning returns, it will also be useful to identify strategies to adapt classroom GIS for in-class learning. With this proposed new pedagogy for teaching and learning with GIS, graduating students will be better equipped to understand and to contribute to the ubiquitous presence of geomedia in our society.

BIBLIOGRAPHY

Albrecht, Jochen. *Key Concepts and Techniques in GIS*. London: SAGE Publishing, 2007.

Battista, Geoffrey A., and Manaugh, Kevin. "Illuminating Spaces in the Classroom with Qualitative GIS." *Journal of Geography in Higher Education* 42, no. 1 (2018): 94–109. https://doi.org/10.1080/03098265.2017.1339267

Canadian Bureau for International Education (CBIE). "International Students in Canada." Canadian Bureau for International Education, 2020. Accessed November 11, 2021. https://cbie.ca/infographic.

Elwood, Sarah, and Wilson Matthew. "Critical GIS Pedagogies Beyond Week 10: Ethics." *International Journal of Geographical Information Science* 31, no. 10 (2017): 2098–116. https://doi.org/10.1080/13658816.2017.1334892

ESRI. "2021 ArcGIS StoryMaps Challenge." 2021. Accessed November 11, 2021. https://www.esri.com/en-us/arcgis/products/arcgis-storymaps/contest/overview.

Government of Manitoba (GoM). "Province Is Building a System That Puts Students First, Gives Educators the Tools They Need and Has a Greater Voice for Parents." Government of Manitoba, 2021. Accessed March 15, 2021. https://news.gov.mb.ca/news/index.html?item=51001.

Hadziristic, Tea. "The State of Digital Literacy in Canada." The Brookfield Institute, 2017. Accessed November 11, 2021. https://brookfieldinstitute.ca/the-state-of-digital-literacy-a-literature-review.

Huynh, Niem T., Bob Sharpe, Chris Charman, Jean Tong, and Iain Greensmith. "Canada: Teaching Geography through Geotechnology Across a Decentralized Curriculum Landscape." In *International Perspectives on Teaching and Learning with GIS in Secondary Schools*, edited by Andrew J. Milson, Ali Demirci and Joseph J. Kerski, 37–47. Arlington, Texas: Springer, 2011.

Jensen, T. P. "Denmark: Early Adoption and Continued Progress of GIS for Education." In *International Perspectives on Teaching and Learning with GIS in Secondary Schools*, edited by Andrew J. Milson, Ali Demirci and Joseph J. Kerski, 73–82. Arlington, Texas: Springer, 2011.

Kerski, Joseph J., and R. Baker Thomas. "Collecting Geo-Data to Support Classroom Field Studies." In *Learning and Teaching with Geomedia*, edited by Thomas Jekel, Eric Sanchez, Inga Gryl, Caroline Juneau-Sion, and John Lyon, 59–69. Newcastle upon Tyne: Cambridge Scholars Publishing, 2014.

Pavlovskaya, Marianna. "Critical GIS as a Tool for Social Transformation." *The Canadian Geographer* 62, no. 1 (2018): 40–54. https://doi.org/10.1111/cag.12438

Schrumm, Andrew. "The Future of Post-Secondary Education: On Campus, Online and On Demand." (2020). Accessed November 11, 2021. https://thoughtleadership.rbc.com/the-future-of-post-secondary-education-on-campus-online-and-on-demand.

Siegmund, Alexander, Daniel Volz, and Kathrin Schulman. "GIS in the Classroom: Challenges and Chances for Geography Teachers in Germany." HERODOT Working Conference, Stockholm, Sweden, 2007.

Terry, Mark. *The Geo-Doc: Geomedia, Documentary Film, and Social Change.* London: Palgrave Macmillan, 2020.

———."EU/ENVS Introduction to Environmental Documentaries." York University, 2021. Accessed May 10, 2021. https://markjterry.com/wp-content/uploads/2021/10/ENVS-1010-S21-May-8-2021.pdf.

———."The Geo-Doc: Enhancing Environmental Education through Geomedia." Association for the Study of Literature and the Environment, 2021. Accessed November 11. 2020. https://www.asle.org/features/the-geo-doc-enhancing-environmental-education-through-geomedia.

———. "The Youth Climate Report GIS Map Project." (2021). Accessed November 11, 2020. https://youthclimatereport.org.

Conclusion

The contributions to this anthology all address the various ways GIS is being used by environmental humanists. They serve as an emphatic response to early criticism that GIS simply has the capacity to gather large amounts of data and that "no treatment of ethical or political issues" (Gregory and Geddes 2014) are possible. The chapters also counter the claim that says that GIS "tends to be limited to certain disciplines, such as Historical Geography, (and) is largely quantitatively based, analyzing census data, for example" (Ell 2010). In all instances, this book examines how the humanities in general, and the environmental humanities in particular, have introduced interdisciplinary research and perspectives to GIS mapping projects.

The common thread in all these new approaches to collecting and representing GIS data is communications. Users of GIS projects with environmental humanities frameworks and content now have a database rich in multimedia components, hyperlinks, digital attachments, and supplementary research embedded in each pin thus creating databases within databases. Having all this interdisciplinary information available in one digital space accelerates the understanding of environmental science and related global issues to scholars, researchers, media, government officials, and perhaps, most importantly, as we seek to end the climate change crisis, the international environmental policymaker. Seen through the lenses of moral and ethical considerations afforded by the environmental humanities, these new GIS projects allow the policymaker to refine laws that recognize and acknowledge underrepresented communities and those most severely impacted by climate change.

We have seen how citizen scientists are now contributing to GIS databases on a global scale further democratizing the process of harvesting information beyond the hallowed halls of academia. Young people and Indigenous people throughout the world are telling stories of climate research, impacts, and solutions in their communities through visual media such as film and photography in GIS projects like the Youth Climate Report (Terry), the Planetary Health Film Lab (Kornberg), and drone mapping in Australia (Jones) helping establish a direct line of communication to the environmental policymakers of the United Nations.

The co-editor and author shows us how digital maps can be effective storytelling tools and provides a valuable framework for constructing narratives in this new medium that tell stories of our relationship to our environments. The storytelling affordance of the new GIS extends itself to the classroom to become a new instrument of instruction for teachers as well as a new platform for projects for students. In chapters contributed by Long, Terry, Tate, and Shehwar new pedagogies for GIS are introduced and in each examination we see the environmental humanities enriching the stories with valuable interdisciplinary research and data in the classroom.

As a knowledge collection and delivery system, GIS is aggressively growing in directions that provide new information and data explicitly through content added and implicitly through content discovered. The unique affordances of analyzing uploaded data temporally and spatially yield findings not inherent in the original data units, and when seen through the interdisciplinary lenses examined in this book, a humanist approach emerges that gives life to cold figures and helps direct progressive new environmental policy for all.

So where does GIS go from here? As digital technologies continue to evolve, so too will their presence in GIS projects. Live videos feeds of places around the world as diverse as eagle nests to downtown urban centers can now be accessed when they are geo-located on digital maps. We can not only see life unfold in real time from the comfort of our own homes, but we can now open a communications channel to the world that will further reduce the size of the global village and introduce a tool to improve all our lives and our relationship to our environments.

REFERENCES

Ell, Paul S. "GIS, e-Science and the Humanities Grid." In *The Spatial Humanities: GIS and the Future of Humanities Scholarship*, edited byDavid J. Bodenhamer, John Corrigan, and Trevor M. Harris. Bloomington, IN: Indiana University Press, 2010.

Gregory, Ian N., and Alistair Geddes. *Toward Spatial Humanities: Historical GIS and Spatial History*. Indiana University Press, 2014.

Index

Page references for figures are italicized.

AchutaRao, Krishna, 84
Aggarwal, Mayank, 84–85
Agarwal, Bina, 88
Alaimo, Stacy, 79, 87–88
Albrecht, Jochen, 178
Alves, Daniel, 115
Antarctica, 11
Anthamatten, Peter, 121
Anthropocene, 83, 109, 134, 143–144
Anthropocene Education Program (AEP), 143
Anthropocene Project, The (TAP), 143
Anthropocene: The Human Epoch, 143–144
ArcGIS, 54, 68, 126, 169, 174,
Arctic, 11, 19–20, 23, 53, 65, 67–69
ArcNews, 66–67, 69–71
Arizona State University, 138–139
Arsenault, Rachel, 56
Arthur, Bill, 127
Assembly of First Nations, 51
Atlantic, The, 146
Autopoiesis, 94–96, 98–100, 103, 105, 107–108

Barad, Karen, 79

Barry, Andrew, 77
Bateson, Gregory, 101–102, 105, 107
Battista, Geoffrey A., 178–179
Bear 71, 99
Bellarine Catchment Network, 32
Bennett, Tony, 77
Bertling, Joy G., 163
Bodenhamer, David J., 93
Booth, Annie, 50
Bosak Keith, 77
Bosse, Amber, 97
Brewer, Cynthia A., 116, 120, 122–123
Briggs, C., 30, 32, 34–35
Briney, Amanda, 145
Brown, P. L., 36
Burroway, Janet, 116–117

Callaghan, Corey T., 18–19
Canada Land Inventory, 9, 19
Canada Wildlife Service, 19
Canadian Bureau for International Education, 180
Canadian Geographic Magazine, 143, 145
Canadian Museum of History, 57
Caquard, Sébastien, 114, 117
Carr, G. W., 42
Carralero, Pamela, 185
Carse, Ashley, 83

Cartwright, William, 114, 117
Casino, Del J., 96–98
Chambers, Kimberlee, 63
Chanana-Nag, Nitya, 85–*86*
Chapin, Mac, 69
Chirnside, Thomas, 37
Christie, M., 36
Citizen Science / Scientist, 12–13, 16–20, 23,183
Citizen Journalism / Journalist, 20, 22
Clark, I. D., 34
Cochrane, Logan, 52
Cole, Steve, 19
Collaboration (or Interpersonal) Competence, 139
Collins, Amy, 121
Colloff, Matthew, 128–129
COP25, 68
COVID-19, 124, 135, 146, 171, 180
Congolaise Industrielle des Bois, 67
Corporate Mapping Project, 99
Courtney, A., 41
Crampton, Jeremy W., 97
Crane, Nicholas, 114
Crawford, Justice, 38–41
Creative Nonfiction (CNF), 116–119, 122, 124, 126
Cree Nation of Mistissini, 49
Cybernetics: Or Control and Communication in the Animal and the Machine, 102

Dahdaleh Institute for Global Health Research, 5, 21, 149, 151, 154, 186, 189
Deconstructing the Colonial View of Wadawurrung Country: Knowledge Drawn from John Wedge's Field Books of 1835–1836, 188
Deleuze, Gilles, 100
Delgamuukw v. British Columbia, 51, 55
Digital Humanities (DH), 93, 100, 109, 149
Digital mapping, 1–2, 97–98, 113–115, 170, 172–173, 177–178, 188

Dixon, D. P., 96
Documentary film, 9–10, 22–23, 99, 106, 150, 159, 170, 187
Dodge, Martin, 96–98, 106
DOGSTAILS, 120, 123
Dombrovskis, Peter, 6
Driedger, Alexander, H., 57
Druick, Zoë, 18
Duxbury, N., 29

Ecocriticism (book), 115
Ecocriticism (term), 2, 4, 6, 87, 93, 115, 169–170
Ecological Footprint, 12–*15*
Education for Sustainable Development (ESD), 134, 136–137
EJAtlas-Global Atlas of Environmental Justice, 69
El-Hadi, Nehal, 116
Environmentology/Enviromentologist, 10–11, 16, 24
eBird, 12, 16–20, 22
Eneas, Bryan, 56
Environmental Careers Organization of Canada (ECO Canada), 141
Environmental Humanities (EH), 1–6, 9–10, 12, 16, 23–24, 93–94, 99–101, 108–110, 133, 169, 173, 183–184
Environmental Justice Atlas, 70
ESRI, 24, 54, 70, 126, 170–171
EU/ENVS 1010: Introduction to Environmental Documentaries (ENVS 1010), 172–173, 175, 179–180

Falconer Al-Hindi, Karen, 78
Fast, Karin, 1
"(Eco)Feminist Visualization," 75–77, 79–81, 85, 88–89
Fink, Daniel, 19
Field, Kenneth, 128
Field Ornithology, Journal of, 18
First Nations, 1, 3–4, 30, 50–58, 64, 69
First Nations Technology Council, 55

Index

Foundation for Environmental Education (FEE), 21
Freeman, Milton M. R., 65–66, 69, 71
Free Solo, 142
Futures Thinking (or Anticipatory) Competence, 139

Gammage, B., 34
Garrard, Greg, 2
Garrett, Bradley L., 164
Garmin, 3
Gaudenzi, Sandra, 95, 98, 100–101, 108
Gawlik, Dale E., 18–19
Ghebreyesus, Tedros Adhanom, 150
Gieseking, Jen Jack, 109
Gillespie, Tarleton, 166
Geo-Doc, 22, 58, 69, 99, 101, 106, 142, 170–180
Geo-Doc: Geomedia, Documentary Film, and Social Change, The, 69, 142
Geographic Information System (GIS), 22, 26, 30, 63, 75, 106, 113, 115, 149–151, 169, 171
Geomedia, 1–3, 5–6, 22, 53, 67–69, 133–134, 138, 141–147, 169, 171, 175, 180
George Brown College of Applied Arts and Technology, 134, 141, 187
Gladwin, Derek, 94
Global Footprint Network, 12–14, 99
Global Positioning System (GPS), 3, 53, 55, 114–115, 175
Google Maps, 3, 16, 159, 173, 175
Google My Maps, 12, 16, 20, 97, 173, 175–176
Gott, B., 32
Grayling, A.C., 116
Greenberg, Jonathan, 67
Greenwood, David A., 159–160
Gregory, Ian N., 108, 183
Grizzly Man, 115
Gruben, Kikoak, 68
Gruenewald, David A., 155, 160, 163
Gryl, Inga, 142

Guattari, Felix, 103, 105, 108
Guterres, Antonio, 117, 135

Hadziristic, Tea, 172
Haldimand Treaty, 70
Hankins, Katherine, 97
Hanna, Stephen P., 96–98
Happening to Us, 68
Harley, J. B., 96, 105–106
Harris, Johnny, 145
Harris, Leila, 95, 97
Hasson, Shabeh, 84
Hayles, N. Katherine, 102
Hazen, Helen, 95, 97
Hemisphere Design, 30
Heritage Victoria, 37
Herries, Jim, 24
Herrera-Bevan, Ezekiel, 68
Herron, Murray, 185
Hewson, Michael, 186
Hinden, Adam, 71
Hirsch, Brett D., 149
Holdgate, G. R., 32
hooks, bell, 163
Hougham, R. Justin, 159–160
Hughes, Christopher, 124
Huynh, Niem T., 172

Indigenous people, 4, 23, 50, 55–56, 58, 63–65, 67, 69–71, 126, 183
Inuit Land Use and Occupancy Project, The, 65–66
Intergovernmental Committee on Surveying and Mapping (ICSM) (Australia), 119
Ivakhiv, Adrian J., 95, 103–105

Jang, Trevor, 51
Jensen, T. P., 171
Jones, David S., 186
Jones, J.P., 96
Joyce, Patrick, 77

Kabuga, Charles, 152
Kannabiran, Kalpana, 79

Kapoor, Aditi, 81, *86*
Keller, Evelyn, 78
Kelly, Lindsay, 50
Kelly, Lynne, 127
Kerski, Joseph J., 171–172
Kimmerer, Robin Wall, 165
Kitchin, Rob, 97
Knowles, Malcolm, 152–153,
Koncan, Alfonz, 68–69
Konecny, Gottfried, 113
Kornberg, Netta, 156, 183, 186
Kwan, Mei-Po, 75

Ladino, Jennifer K., 109
Lahiri-Dutt, Kuntala, 76, 80–81
Lancet, The, 154
Lansing, Stephen J., 83
Latour, Bruno, 79
Law, Michael, 121
Leadbeater, Richard, 24
Learning and Teaching with Geomedia, 142
Lettvin, John, 102
Levin, John, 100
Levine, Richard C., 84
Lickers, Kathleen, 52
Lin, Maya, 94, 97, 99, 107
Living Documentary, The (term), 95, 99–101, 177
Long, Michael John, 187
Longley, Paul A., 3, 123
Lukacs, Martin, 53
Lundy, Kathleen Gould, 156
Lunt, I., 32

Mackenzie, Scott, 20
Macquarie Atlas of Indigenous Australia, 127
Maneice, Damon, 145
Manaugh, Kevin, 178–179
Manufactured Landscapes, 143
Margetts, V., 30
Martusewicz, Rebecca A., 161
Matawa First Nations, 50
Mathur, Roshni, 84

Mattingly, David, 78
Maturana, H. R., 94–95, 99–100, 102–103
Mayhew, Susan, 119
Mbendjele People, 66–67, 69–70
McGetrick, Jennifer, 66–67, 69
McGregor, Deborah, 53, 58, 150
McLafferty, Sara, 78
Mehta, Avantika, 80
Mercator, Gerardus, 114, 128
Meru, 142
Millar, David, 9
Mohammed, Omar, 146
Monture, Patricia, 58
Moola, Faisal, 51
Moore, Michael Grahame, 153
Morphy, Frances, 127
Mukherjee, Falguni, 97
Murujuga Aboriginal Corporation, 29

Naked World, The, 87
Narayan, Uma, 78–79
Nash, Kate, 101
National Aeronautics and Space Administration (NASA), 19
National Film Board of Canada, 9
National Geographic, 120, 170
Neale, Margo, 127
NextGen Video Challenge, 21
Nicholson, M., 27–28, 30, 33–34
Nicolosi, Emily, 94, 106
Night Moves, 142

Olson, Randy, 119
Open Street Maps, 97

Parks Victoria, 29
Participatory GIS (PGIS), 5, 49, 64, 71, 95, 97, 106
Pascoe, B., 35
Pattnaik, Itishree, 88
Pavlovskaya, Marianna, 77, 97, 149, 170
Pearce, Margaret Wickens, 123

Pedagogy (ies), 2, 5, 146, 149, 151–153, 155–156, 159–161–166, 169–171, 180, 184
Pedagogy of Place, 149, 153, 155–156, 159–166
Perl, Sondra, 116, 118, 120
Pickles, Thomas, 113
Pierre, Dennis, 68
Planetary Health Film Lab, 5, 21–22, 149, 151, 154, 156–157, 159, 161, 164, 183, 189
Poole, P., 29
Powell, B., 32
Pritchard, L., 38
Promise of Infrastructure, The, 83

Queiroz, Ana, 115

Rambaldi, Giacomo, 49–50
Red Deal: Indigenous Action to Save Our Earth, The, 64
Red List of Ecosystems, 99
Red List of Threatened Species, 99
Red Nation, The, 64–65
Reuschel, Anne–Kathrin Weber, 123
Rickards, Nathan, 84–85
Ridanpaa, Juha, 115
Rigby, Colin Wayne, 117
Roberts, H., 35, 37, 39
Roös, Phillip, 187
Rose, Deborah Bird, 1, 27–28, 101
Roth, Robin, 52
Royal Canadian Geographical Society, 143
Rozhon, Jon, 56
Ryan, Lyndall, 126
Ryan, Susan, 188

Salleh, Ariel, 89
Sapelli (software), 66–67
Schrumm, Andrew, 179
Schulman, Kathrin, 172
Schultz, David M., 118
Schultz, Julianne, 129
Schultz, Nikolaj, 79

Schwartz, Mimi, 116, 118, 120
Shadian, Jessica, 64
Shah, Anil, 80
Shehwar, Shereen, 174, 184, 188
Siegmund, Alexander, 172
Skelton, Norm W., 50
Smith, Gregory A., 156, 163
Smith, Jane M., 51
Social Sciences and Humanities Research Council of Canada (SSHRC), 149
Somerville, Margaret J., 165
Spatial Humanities, 170
Stancombe, G. H., 38–39, 41–42
Steffen, Will, 143–144
Steinitz, C., 28
Stenport, Anna Westerstahl, 20
StoryMaps, 170
Strategic Thinking (or Action-Oriented) Competence, 139
Strobl, Josef, 133
Sui, Daniel, 123
Sullivan, Brian L., 17–18
Sustainable Development Goals (SDG), 5, 21–22, 134–136, 170, 177, 179
Sutton, Jeff, 49
Sword, Helen, 118
Systems Thinking Competence, 138

Talmage, James, 145
Tapisirat People, 67
Tate, Erik, 174, 184, 188
Taylor, Joanna E., 115, 123
Terry, Mark, 58, 67–69, 94, 97, 99, 106, 113, 141–142, 144, 151, 169–170, 173, 183–184, 189
Thomas, Leah, 115
Threadgold, H., 30
Three Ecologies, The, 103–105, 110
Tigresa, La, 87
Tobler, Walter, 3
TODALSIGS, 120
Tournier, D., 32
Tomlinson, Roger, 9, 96, 113
Townsend, Justine, 51

Traditional Ecological Knowledge (TEK), 55, 59
Tsering, Jigme Lhamo, 189
Tsilhqot'in Nation v. British Columbia, 51–53
Twister, 142

UNESCO Chair in Reorienting Education Towards Sustainability, 135–137, 189
United Nations, 5, 20–22, 63–64, 106, 117, 134–135, 137, 151, 159, 170, 173, 177, 184
United Nations Decade of Education for Sustainable Development, 137
United Nations Educational, Scientific and Cultural Organization (UNESCO), 29, 134–137, 170, 189
United Nations Environment Programme (UNEP), 117
United Nations Framework Convention on Climate Change (UNFCCC), 20, 151, 170, 189
University of Calgary, 99
University of Newcastle, 126
Upadhyay, Bhawana, 80

Values Thinking (or Normative) Competence, 139
Varela, F.J., 94–95, 99–100, 103
Volz, Daniel, 172

Walden, Zoey, 56
Water for All, 81–82, 85
Watermark, 143
Wedge, John Helder, 38–40
Weik, Arnim, 135, 138–139
What is Missing?, 94, 99, 107–108
Whalen, Jake, 51–53
Whitmee, Sarah, 154
Wiener, Norbert, 102
Wilfrid Laurier University, 21
Wilkie, B., 32
Wilson, Kory, 53, 58
World News Day, 21

York University, x, 6, 12–15, 21, 136, 149, 151, 157, 172–173, 180, 187–189
York University's English Language Institute (YUELI), 180
Young Lives Research Laboratory, 149
Young Reporters for the Environment (YRE), 21
Yousafzai, Malala, 135
Youth Climate Report (YCR), 12, 20, 21–23, 58, 68, 70, 94, 99, 101, 106, 142, 149, 151, 157, 159–160, 164–165, 170, 183, 189
YouTube, 145, 159, 173, 175
Yusoff, Kathryn, 79, 88

Zhu, Xuan, 96
Zwarteveen, Margreet, 80–81, 85

About the Contributors

PAMELA CARRALERO

Pamela Carralero is professor of environmental humanities at Kettering University, Flint, Michigan, USA.

Pamela Carralero teaches environmental humanities at Kettering University and works as an environmental resilience consultant. She has recently consulted for the University of Southampton's project, Building REsearch Capacity for sustainable water and food security In drylands of sub-Saharan Africa (BRECcIA). She holds a PhD in theory and cultural studies from Purdue University (2019) and an MSc in literature and transatlanticism from the University of Edinburgh.

MURRAY HERRON

Murray Herron PhD, American Planning Association (Member); Canadian Institute of Planning (Member).

Murray possesses an in-depth knowledge of strategic land use planning and urban design, gained as a local government strategic land use planner in Australia and as a graduate from Deakin, RMIT, Swinburne, Melbourne, and BCIT universities. In his PhD titled "Climate Change Resilience and Development Growth of Two Australian Cities" (2016), his research offered internationally relevant applied findings that have been presented at and are applicable for national and international land use planning, modeling, and visual scenario-making and assessment application.

MICHAEL HEWSON (EDITOR)

Michael Hewson PhD, senior lecturer in geography at Central Queensland University.

Dr. Michael Hewson is an environmental geographer, whose research revolves around Earth observation using spatial science techniques. Research projects have included threatened species habitat health monitoring; numerical weather modeling of aerosol transport; land surface temperature mapping; and land-use change mapping. The prime motivation for Michael's academic interest is the Earth system science imperative—the anthropogenic impact on the interconnectedness of nature.

https://staff-profiles.cqu.edu.au/home/view/139

DAVID S. JONES

David S. Jones PhD, MPIA, FAILA Reg Land Arch, M.ICOMOS

David is oversighting Strategic Planning and Urban Design at the Wadawurrung Traditional Owners Aboriginal Corporation, and is also an adjunct professor with Griffith University's Cities Research Institute, and a professor (research) with Monash University's Monash Indigenous Studies Centre. With academic and professional qualifications in urban planning, architecture, and cultural heritage, he has taught, researched, and published extensively across these disciplines areas in the last thirty years, including Indigenous knowledge systems.

NETTA KORNBERG

Netta Kornberg, Knowledge Exchange Lead at Ontario Harm Reduction Network, Toronto, Canada

Netta Kornberg, MPhil, is the Knowledge Exchange Lead at the Ontario Harm Reduction Network and was previously the Knowledge Dissemination Strategist at the Dahdaleh Institute for Global Health Research. She has worked internationally in adult education, public health, and the arts.

MICHAEL JOHN LONG

Michael John Long, MES, LLM; contract faculty professor, School of Liberal Arts & Sciences; George Brown College of Applied Arts and Technology
Michael teaches geography and sustainability as a contract faculty professor at George Brown College of Applied Arts and Technology (GBC). At GBC, Michael is also the Faculty Advisor to the student Sustainability Squad, and a member of the Tommy Douglas Institute. In the community, Michael works on the programming team at Canada's largest international environmental film festival, Planet in Focus. Michael has published book chapters, articles, and interviews about the environment and documentary film, and notably with Canada's premier documentary culture magazine, *Point of View*. He is the 2021 Awards of Excellence recipient for "Outstanding Post-Secondary Educator" from the Canadian Network for Environmental Education and Communication (EECOM). Michael studied at York University and Osgoode Hall Law School where he completed the master of environmental studies and master of laws, respectively.

PHILLIP ROÖS

Phillip Roös PhD, MPIA; RAIA; AILA
Phillip is an ecological systems–inspired architect, designer, planner, and author for regenerative architecture, sustainable urban developments, infrastructure, and landscapes. He is currently the director of the Live+Smart Research Laboratory at the School of Architecture and Built Environment, Deakin University. His research interests are centered on the human-nature relationship and the identification of optimized design processes based on a regenerative-adaptive pattern language system. In professional practice he has been working as a design professional for over thirty years on an extensive range of small- to large-scale projects in Europe, Africa, and Australasia. His work spans biophilic design, architecture, urban design and planning, landscape architecture, environmental design, Indigenous knowledge systems, teaching and research, as well as writings and ecological art. His work is influenced by whole systems thinking and his application of environmental design is closely related to the ordering of the large-scale aspects of the environment by means of design and ecological planning.

SUSAN RYAN

Susan Ryan MPlan (Prof), Deakin; MPIA; Aust.ICOMOS (Assoc.); AHA (Member)

Susan is the principal planner of People Earth Design, after approximately eleven years within local government, specifically Arts and Cultural Services in Victorian's eastern suburbs. She compliments this as a research member for Indigenous knowledge systems for the Live+Smart Research Laboratory, Deakin University. A graduate of a Master of Planning (Professional) with specialization in cultural heritage from Deakin University, she continues her discovery and understanding of Country through as doctoral candidate researching *Deconstructing the Colonial View of Wadawurrung Country: Knowledge Drawn from John Wedge's Field Books of 1835–1836.*

SHAHREEN SHEHWAR

Shahreen Shehwar, PhD candidate, Faculty of Environmental and Urban Change, York University, Toronto, Canada

Shahreen's research interests lie in the areas of urban planning, land-use planning, and Indigenous jurisdiction. Her main research interest centers on Indigenous populations' differential access to space in Canada and the factors behind it. She is currently exploring opportunities to use her research and analytical skills to contribute professionally and academically in the areas of Canadian environmental policy.

ERIK TATE

Erik Tate, PhD candidate, Department of Humanities, York University, Toronto, Canada

Erik positions his work in the environmental humanities, and his research interests include environmental justice, ecosophy, and the intersection between technology, semiotics, and ecology. He is currently working on an interdisciplinary dissertation provisionally titled "Semiotic Ecologies of Perception, Space, and Interactivity: An Ecosemiotic Approach to Interactive Digital Mapping Projects."

MARK TERRY (EDITOR)

Mark Terry, PhD, FRSC; adjunct professor, Faculty of Environmental and Urban Change, York University, and Department of English and Film, Wilfrid Laurier University; Associate to the UNESCO Chair in Reorienting Education towards Sustainability; Executive Director, UNFCCC's Yputh Climate Report.

Mark Terry teaches courses on geomedia, GIS, ecocinema, documentary film, and climate change. He is also one of the project leads of the Planetary Health Film Lab at the Dahdaleh Institute for Global Health Research.

Profile: https://markterry.academia.edu

JIGME LHAMO TSERING

Jigme Lhamo Tsering, MES candidate, Faculty of Environmental and Urban Change, York University, Toronto, Canada

Jigme's research interests are centered around the intersection of environmental justice, GIS accessibility, and water governance with a specific focus on the Indigenous narratives of the Himalayan people. This interest stemmed from a research paper she published in the University of Toronto Mississauga's *Young Anthropology Journal* on the water bottling industry and its extractive practices across the Tibetan plateau. Through academia, Jigme aims to examine policies and contribute to the development of sustainable and inclusive water policy.